对谈
dialogues

atelier ruelle
街巷工作室 设计作品专辑

Green Vision 绿色观点·景观设计师作品系列

本系列图书为法国亦西文化公司(ICI Consultants/ICI Interface)的原创作品，原版为法英文双语版。
This series of works is created by ICI Consultants/ICI Interface, in an original French-English bilingual version.

法国亦西文化 ICI Consultants 策划编辑

总企划 Direction：简嘉玲 Chia-Ling CHIEN
协调编辑 Editorial Coordination：尼古拉·布里左 Nicolas BRIZAULT
英文翻译 English Translation：爱玛·望多尔 Emma VANDORE
英文校阅 English Proofreading：艾莉森·库里佛尔 Alison CULLIFORD
中文翻译 Chinese Translation：王美文 Mei-Wen WANG
中文校阅 Chinese Proofreading：简嘉玲 Chia-Ling CHIEN
版式设计 Graphic Design：维建·诺黑 Wijane NOREE
排版 Layout：卡琳·德拉梅宗 Karine de La MAISON

绿色观点·景观设计师作品系列

green vision

对谈
dialogues

atelier ruelle
gérard pénot & associés

街巷工作室 设计作品专辑

弗蕾德里克·德·格拉维廉 撰文

辽宁科学技术出版社

contents 目 录

前言 – 彼此互应
introduction – living together … 006

结伴同行
companionship … 008

享受乐趣
pleasures … 056

自家楼下
on our doorstep … 098

亲自栽种
green fingers … 118

方案索引
projects index … 130

工作室简介
biography … 134

版权说明
credits … 135

致谢
contributions … 135

introduction

前言

living together
彼此互应

城市生活的乐趣,来自于城市空间组成元素的质量,以及它们彼此之间建立起的关系。这也便是城市整治的目的,在物体与空间之间建立起和谐的联系,使它们并非彼此排置并列而已,而是能够相互应和,摒弃粗俗与松散,让人获得美感与细致这些有时难以言喻的感受。

共通语汇

塑造公共空间是这项挑战的核心,无论是就它带来的生活质量而言,或是为了其他更远大的抱负:成为一个地方政府进行城市政策的工具。为了扮演这个重要角色,公共空间必须要能够在时间中更迭转换,成为永续发展过程的一部分,与所有的街区息息相关……街巷工作室在圣纳泽尔的城市项目便应用了取自巴塞罗那的城规经验,唯有城市整治方案才能使道路呈现社会与文化的尺度,并与整体协调一致。

公共空间这个词本身便已说明:这是属于所有人的空间。为了创造"公民作品",必须将所有相关的人们——政治家、居民、管理者……结合在一起,并且需要塑造一种共通语汇来表达场所精神。这个语汇结合了空间组构线条的严谨性(通过坚固的材质、人行道、组装细节等来呈现)和让人感到舒适的流畅性,不仅不把空间界定的太清楚,同时也遵循规范行事。规范是和谐的基础,因为它界定了如何共存的方式。

创造生命力

处理公共空间会触及地面楼层的问题,结合这两者所面临的挑战是:如何创造生命力与建立关系,如何与他人有所接触。思索行人穿梭来去的都市,便是想象如何从一个空间到另一个空间,考虑其途径的安排,思索建筑体量之间的穿透流通,研究入口大厅或一道隔墙的透明性来丰富建筑与街道的关系。

The pleasure of city living comes from the quality of the objects that make up the urban environment and the relationship between them: ultimately, this is what urban planning hopes to achieve. It is about creating relations between spaces and objects, not by putting one next to another, but by making them interact with each other. It is about safeguarding against vulgarity and neglect, and finding beauty and delicacy – sensations that can be felt without always having to be explained.

A common language

This challenge is at the heart of any work involving public space, both in terms of the quality of life it brings and the wider ambition of being a tool for the urban policy of a local authority. For this to be achieved, however, it should be seen as part of a wider, long-term development that affects all neighbourhoods. This lesson, drawn from Barcelona's urban planning, was employed in France's Saint-Nazaire, where social and cultural dimensions at street level were part of a coherent whole only possible in urban projects.

It is what is says: public space, for all to enjoy. For a real "community project", different groups must work together: politicians, residents, managers... They should create a common language, which helps to create a common state of mind. This language should combine strong lines, expressed in the solidity of materials, sidewalks and construction details, and the fluidity that brings a feeling of amenity. At the same time, it should try to avoid overly defining the spaces, whilst respecting standards that have already been laid down. Such norms define how to live together, allowing people to get along.

The creation of a living space

The issue of public space encompasses everything at ground level, which is where public and private worlds meet. Both meet an essential challenge – the creation of a living space, of relationships, and being in contact with others. Think of the city from the perspective of the passer-by: it is about moving from one space to another, the pathways and porousness formed between the masses, the transparency of a lobby or an enclosure that enriches the relationship between buildings and the street.

工作步骤

街巷工作室的工作步骤由以下几个原则组成：

- 构思让人一目了然的场所：容易解读的空间会让人感到舒适自在，而能够在其间轻易地辨认方向是让人产生安全感不可或缺的条件。

- 简约行事：这个严谨的训练能以最低的成本争取最大的使用面积。这代表着避免让空间具有过于特定的使用功能，以保证空间的流畅性。这也意味着对空间持久性的保障，使其能够轻易地获得新的整治，而无需过度干扰或扭曲既存的一切。

- 预先考虑保养维护事宜，并与项目管理者和市政单位沟通：空间的质量必须透过妥善的管理来维持。

- 结合新与旧，将空间整治设想成随着时间而层层叠合的发展方式，未来的规划将奠基于今日的成果。

- 特别关注方案的实施过程，以发挥所使用材料的最佳效果，如地面、照明、植被、与天空的关系。

一场舞蹈

城市整治项目的精神代表一种永续持久的工作，因此必须建立时间层面上的策略和施行时的务实条件。

这样的方式意味着承担起一种和传统的城规任务全然不同的责任。任务的时间必须充裕，以便研究由不确定因素造成的各种演变情况，并得以和各种不同参与者交涉。和居民、地方机构、开发商及其委任建筑师、政府部门及业主的协商是项目进程中攸关成败的工具。身为城市项目的规划设计者要能够将这些交流视为一种乐趣，即使有冲突纷争，也必须能以温和的态度进行沟通。

城市规划变成一场各种人物参与演出的舞蹈，而每个人对城市的看法各有不同。城市规划师扮演了一个矛盾的角色：他既代表着整体，也呈现属于自我的创作；他在说出"我存在"的同时也表达了："我要你们存在。"

The approach

Several principles inform the Atelier Ruelle approach.

- Imagine places that are effortlessly perceived. That they are easily understandable contributes to their comfort. And it is essential to get your bearings easily to feel safe in the city.

- Keep it simple. This Spartan exercise allows us to make the most out of the space at the lowest cost. It is about ensuring a fluidity, avoiding overly specialized designations. And also guaranteeing sustainability: allowing easy interventions that don't disrupt or distort what exists.

- Anticipate upkeep, and discuss this with management and city services. The sustainable quality of spaces depends on their maintenance.

- Combine the old and the new. Think of development as a set of multiple layers deposited over time. What is built today will be built upon in the future.

- Pay close attention to the implementation in order to make the best use of materials such as pavements, lighting, planting, and the perception of the sky.

A dance

The spirit of the urban project requires a long term approach, requiring temporal strategies and pragmatic conditions for completion.

Such an approach requires a commitment that is incommensurate with the classic mission of the planner. Missions must allow planners to take their time, to study the evolving scenarios according to uncertainties, to negotiate with multiple actors. Negotiation is a crucial tool in the process – with residents, local authority services, developers and their architects, institutions and landowners… Being an urban contractor requires being able to take pleasure in these discussions, to be able to take it in one's stride even when there are conflicts.

In this way, the planning process becomes a dance between people who each think of the city in a different way. The planner has a paradoxical role in this: representing the community while at the same time expressing something of himself, saying at the same time "I exist" and "I want you to exist."

companionship 结伴同行

对一块基地进行规划既与人也与时间相关。街巷工作室相当热衷于这种必须持续多年的工作，借此熟悉一个都市以及其民意代表和部门机构。都市方案所提出的构想往往是一个渐进的实施和发展过程，透过与不同角色的沟通交流而日益丰富，建立了一个共通的文化。

历时长久的城市需要其不同决策者在时间的长河里保持一定的发展逻辑，街巷工作室针对这个持久性的层面发展出一些相对策略，包括：选择从何处快速着手，以便赢得信誉（特别是面对敏感街区的民众与投资商的时候），并激发继续进行改造计划的意愿；由城市外围与边缘开始，进行环状或点状渗透的规划工作；避免不确定因素与障碍对方案造成影响，为其提供既有力又具有弹性的施行条件，不执著于单一而激进的解决方式；不冒险陷入僵局，但绝不轻言放弃。

街巷工作室的工作方法奠基于对下列事项仔细的分析评断：原有物理环境与社会环境的多样性；方案的目标与意图。为了防止规划方案遭到弃置的命运，必须避免提出会造成两极化（赞成/反对）的建议，也必须避开会引起正面冲突的方法（造成合作者筋疲力尽或资金枯竭的风险）。

To intervene on a site is as much a human question as one of duration. The Ruelle team like their work to continue for years, allowing them to get to know a city, its elected officials and its services. Conceived as a process, the project develops by progressive enrichment through exchanges with its actors. It establishes a common culture.

The long view that must be taken when working in any city environment requires a lasting consistency from decision makers, a concern for permanence towards which Atelier Ruelle develops strategies. It is about choosing where to intervene quickly to gain credibility (essential vis-à-vis local communities in underprivileged neighbourhoods as well as investors) and to inspire a desire to carry on, to take a roundabout approach or to infiltrate, to start from the periphery and work in from the edges. To keep the project going despite uncertainties and obstacles, whilst ensuring that the conditions for completion are both strong and flexible, without being attached to a unique or radical solution. Never to risk stalemate, but never to give up.

Atelier Ruelle's approach is based on an exacting diagnosis: the plurality of what exists, both physical as well as social, and the goals and intentions of projects. Faced with the risk of abandonment, proposals should avoid binary situations (for/against) and head-on approaches (which risk exhausting partners and funding).

法国 圣纳泽尔 / 自1988年起

building a common culture
造 就 共 通 文 化

自1988年以来在圣纳泽尔市的经验一直是个非凡的例子。在一段为期长久的时间里，由城市规划师组成的团队陪伴着一个城市、其市长和技术服务部门一起进行了许多不同的方案。这个结伴同行的过程逐渐形成一种共通的文化。对街巷工作室而言，和日后负责落实方案与进行维护工作的人员一起构思整治项目确实是必需的，而不同的工作单位一起共享方案的文化也是相当重要的。

参与这个经验的人员都是开路先锋，他们创造了圣纳泽尔风格，也创立了几个游戏规则，成为随后城市规划领域里的模范准则：1980年代末期，法国境内试行的几个公共空间整治都是从巴塞罗那的城市规划中汲取灵感。

Since 1988, the experience of Saint-Nazaire has been exceptional. A team of planners has served a city, its mayor and its technical services over a long period and across a variety of projects. This companionship has gradually forged a common culture. According to Atelier Ruelle, it is essential to design facilities with those who will make them happen and maintain them, just as it is essential that the services share a project culture.

The original pioneers in this adventure invented the Saint-Nazaire style. In doing so, they also invented some rules that have subsequently become a model in terms of urban design: in the late 1980s, the early developments of public spaces were being tested in France, greatly inspired by Barcelona.

城市乐趣

圣纳泽尔把赌注放在城市建设上，企图为这个面临严重经济危机的都市带来经济复苏。公共空间的整治扮演了首要的角色，目标是重新树立圣纳泽尔的城市形象，这并非透过行销手法来为城市进行宣传，而是让大家重新感受城市生活的乐趣，让居民以身为圣纳泽尔市民为荣。

城市生活的乐趣首先与市中心有关，当时的市中心已日渐丧失了活力，因此必须被优先考虑。交由建筑师克劳德·瓦斯科尼设计的"船舶"商业中心增加了市中心的密度，并发展了新的商机。公共空间的规划工作伴随着这个项目开展，将一条国道转化为大街，从而建构出一条充满吸引力的城市路线。

继这个具有决定性的项目（1987-1988）之后，市中心接而开始朝着火车站与港口区扩张（1991-1993）。在"都市-港口"城市项目（1995-2002）当中，加泰罗尼亚的都市规划师曼努埃尔·德·索拉莫列斯引导城市转向其往昔的潜艇基地上，进而向海边延伸。"都市-港口"规划项目的第二阶段（2003-2010）更确认了这个方向：让城市朝着港口寻回过去的历史。

圣纳泽尔这个城市原本不具有都市建设上的声誉，却因推展"都市-港口"项目而声名远播，城市自此之后仍然继续在蜕变中。

The pleasure of city life

Saint-Nazaire is putting its money on emphasising the urban in order to revive a city suffering a major economic crisis. Work on public space plays a leading role in redressing the image of Saint-Nazaire. This does not mean city branding, but instead reconnecting the inhabitants with the pleasure of city life, restoring pride in being from Saint-Nazaire.

This pleasure of city life approach began with the centre, a priority issue because it had lost its vitality. The "Paquebot" shopping centre by architect Claude Vasconi gave it a density and a new commercial offer. Work on public spaces was part of this programme, transforming a trunk road into an avenue and starting to create an attractive urban route.

This intervention at the crossroads (1987-1988) was followed by the first expansion of the heart of the city to the station and to the port (1991-1993). With the City-Port project (1995-2002), Catalan urban planner Manuel de Solà-Morales moved the city's focus back towards its submarine base and the sea beyond. The second phase of City-Port (2003-2010) has confirmed this city trend of reconnecting with its history via the port.

A city without an urban reputation, Saint-Nazaire suddenly attracted interest during the City-Port project, and is continuing its transformation.

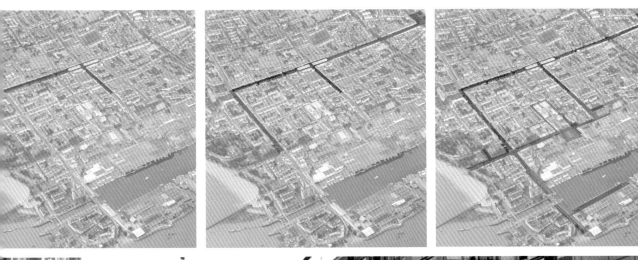

在圣纳泽尔逐步进行的项目,最上方简图由左到右:
1987-1988 城市中心 / 1991-1993 城市中心朝着火车站与港口区进行初步扩张 / 2005-2012 都市-港口第二阶段

Successive interventions, from left to right:
The city centre, 1987-1988 / First extension of the heart of the city towards the train station to the east and the harbour, 1991-1993 / City-Port Phase 2, 2005-2012

渐进的积累

在"都市-港口2"项目中，街巷工作室延续并全揽之前与曼努埃尔·德·索拉莫列斯合作过程中开始对地理形态的研究，将新的方案融入先前已经进行的方向。毫无疑问地，这是为了尊重最初的设计者，但同时也是基于对城市执行能力的考量，因此需要一个渐进的过程。通过许多不同的项目，街巷工作室陆续为此城市进行着规划与设计，在终究为期不算长的15年历史中，每一步都在进行建设、回收和融合，每一个新完成的行动都为其相应的都市政策带来意义。

Progressive accumulation

As part of City-Port 2, Atelier Ruelle is continuing and refining the survey work started in collaboration with Manuel de Solà-Morales. This approach slips seamlessly into the movement already started, partly of course out of respect for the designer, but also taking into account the city's resources, which require a step-by-step approach. Through the successive interventions that the Atelier has led on various projects, it has been building, recycling and incorporating at each step – and all this in a relatively short time frame of fifteen years. Viewed cumulatively, each action has given meaning to urban policy.

拉丁美洲广场，特别选择原生长于南美洲的植物种类：智利南洋杉、丝兰、布迪椰子

Latin America Square (Place d'Amérique Latine), tree species selected for their origin: Araucaria araucana, Yucca, Butia

17

简约的风格

圣纳泽尔城市空间具有简约的风格：人行道上的沥青路面在雨后泛发着光泽，地面铺砖的连续条带与成排树木的线条强化了空间感，棕榈树令人想起让港口在战前光荣一时的横渡大西洋旅程……

A simple style

Saint-Nazaire style is simple: asphalt pavements that give off light after the rain, continuous strips of paving stones and rows of trees highlight space, while palm trees bring to mind the transatlantic voyages that were the glory of the port in the pre-war years.

上图：文森·奥里奥尔路，改造前后面貌
右图：戴高乐大道

Above: Rue Vincent Auriol before and after
Opposite page: De Gaulle avenue

遍及每个街区

让人们重新热爱都市、向往住在都市，这样的新计划不仅涉及城市空间的使用功能，也关乎城市形象的更新，不仅和市中心街道有关，也包括车站、海滩、社会住宅区……不为项目列分等级，但需要拟定资金支出的策略：有待整治的空间繁多，而所拥有的财力极少，因此必须极力节约，将少量资金延展到最大面积来使用，才能挪出足够的预算来支付标志性场所所需的特殊整治。

In all districts

Reinvesting in the love of the city, the desire to live there, concerns both the lived experience and a change of image. It affects city centre streets as much as the train station, the beach, social housing neighbourhoods... The Atelier's approach eschews hierarchy but prioritises an expenditure strategy: with many spaces requiring attention and little money available it was necessary to economise, to stretch funds to the maximum to ensure funds were available to pay for specific interventions on emblematic places.

布列特里住宅街区　　　　　　　　　　　The residential area Bouletterie

21

属于大众的海滩

街巷工作室也为滨海街区进行规划，这一带因贾克·大地这位名导演在圣·马可海滩所拍摄的"于洛先生的假期"电影而引人注目。就在离拉伯尔不远处的海滩上，一座在其下方安置了餐厅的浮桥式观景平台以朴实的面貌显现出一个大众化海滩的特质。

A beach for the people

Atelier Ruelle is also involved in the seaside neighbourhood, revealed by filmmaker Jacques Tati, who filmed *Les Vacances de Monsieur Hulot*, or *Mr Hulot's Holiday*, on the beach of Saint-Marc. A pontoon gazebo with a restaurant underneath encapsulates all the appeal of a popular beach without frills, which it remains despite its proximity to the smart seaside resort of La Baule.

法国 南特 / 2004-2017

reclaiming places at the margin
争 取 边 缘 场 所

马拉科夫-佩古谢这个城市大计划的长久历程始自2001年的项目研究,方案将持续至2017年,一如长期战斗的过程。这个方案呈现出持久的韧劲和能够避开僵局的必要性策略,同时也表现出对于公家合作方与私人合作方彼此接触的慎密思考,以及和开发商与建筑师的交涉协商。所有这些工作都和公共空间本身的设计一样重要。

这个占地164公顷的项目必须考虑如何转化现存的环境,尤其是具有1700个住房单位的大群体社会住宅区,同时也得建造一个新的混合型街区("欧洲南特计划"就业区、住宅、公共设施、商店),以满足三个层面的需要:基地、街区关系以及地面。项目的目标在于重新让这块土地融入卢瓦尔河谷地,使它与外围环境连接、便于穿越,成为"跟其他地方一样"的街区。方案也同时着重于加强行人的感官感受,发展卓越的空间品质。

The *Grand Projet de Ville* regeneration programme in Malakoff-Pré Gauchet, which started with a definition study in 2001 and will continue until 2017, has been a challenge every step of the way. This project illustrates the need for tenacity and for a strategy that avoids deadlock as well as a delicate approach to the dialogue between private and public entities and negotiation with developers and architects. All this work is as important as the design of the public space.

On this project, with a perimeter of 164 hectares, it was necessary to think about both the transformation of what existed, including a large estate of 1700 housing units, and the creation of a mixed neighbourhood (the Euronantes jobs district, housing, facilities, shops). It had a threefold need: the site, the link between districts, and what is happening on ground level. It is about establishing the territory again as part of the Loire valley, connecting it and crossing it so that it becomes a neighbourhood "just like the others", while also broadening the perception of the pedestrian and cultivating outstanding qualities.

卢瓦尔河畔马拉科夫

这块诞生自卢瓦尔河的土地长期以来便易遭水患，在1960年代以沙土填充，土地范围以卢瓦尔河及圣费利克斯运河界定。由于汽车交通所带来的约束限制，使土地丧失了来自卢瓦尔河的特色，要重新寻回这个特色，意味着必须重新建立起南特和水、河口及其连续不断的自然空间之间具有象征性与历史性的关系。

收复失土

马拉科夫最初的改造计划偏离了重心：尽管位于卢瓦尔河南岸河畔，城市与河流的关系却因为针对交通问题而采取的保护设施而变得抽象。新的方案必须将此情况纳入考量，不能期望车水马龙的大道得以"缓和"（每天有4万往来车辆），因为减少汽车流量是不可能在近期内达成的。

为了弥补这一点，新方案以公共空间和花园来建立和卢瓦尔河的关联，同时在佩古谢区和大群体社会住宅区拆除450个居住单位后，便能利用这些清出的空间来建造靠近河岸的新建筑。

MALAKOFF-SUR-LOIRE

This land born of the Loire, at flood-risk for many years and backfilled with sand in the 1960s, has the river and the Saint-Félix canal as its boundaries. Restoring its Loire valley identity, lost due to the constraints of traffic, means reconnecting with the symbolic and historic link that Nantes has had with water, with the estuary and its chain of natural areas.

The re-conquest

The initial Malakoff project has been driven off course: although located on the Loire facing south, its relationship with the river had become abstract thanks to devices designed to protect against traffic. The project had to accommodate this situation without hope of "traffic calming" on the boulevard (40,000 vehicles per day), because reducing the role of the motorcar is not a foreseeable option in the short term.

To compensate, public spaces and gardens create connections with the Loire, both in the Pré Gauchet neighbourhood and in the housing estate, where the demolition of some 450 dwellings has cleared spaces and allows for new construction near the river.

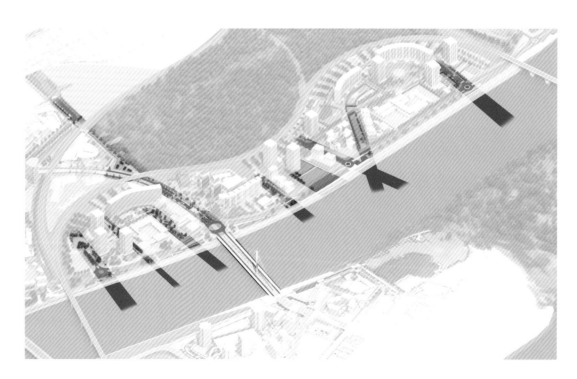

上图：新规划整治的公共空间与花园朝向卢瓦尔河敞开
右图：位于小亚马逊湿地（"Natura 2000 计划"中的保护区）边缘的散步道

Above: Opening of new public spaces and gardens leading onto the Loire
Opposite page: Pedestrian walkway at the edge of the Petite Amazonie, a protected zone under the "Natura 2000" initiative

与卢瓦尔河的天然风貌共存

卢瓦尔河的天然风貌显现在这块土地上。莫沃草原区、德拉罗什公园和小亚马逊湿地是过去洪泛频繁的草原留下的痕迹。德拉罗什公园位于城市边缘，经过整治而变得清晰明显以后，成为了具有实用性的公园，而小亚马逊湿地这个位于市中心的奇特自然空间仍旧属于保护区（为Natura 2000计划的一部分）。

河岸经过重新设计和绿化，期望这些散步道在日后能够将城乡区域的整体自然空间融入其范围中。

Living with the Loire in its natural form

The Loire enters the area in its natural state. The Purple Meadow (Prairie de Mauves), Park de la Roche and the Petite Amazonie are what remains of the historic water meadows. At the edge of the city, the Park de la Roche, cleared up and made more visible, has become an inhabited park, while the Petite Amazonie, an amazing natural area in the heart of the city, remains a protected area (Natura 2000).

The banks have been redesigned and planted, in the hope that these walkways will later be integrated with all natural areas of the city.

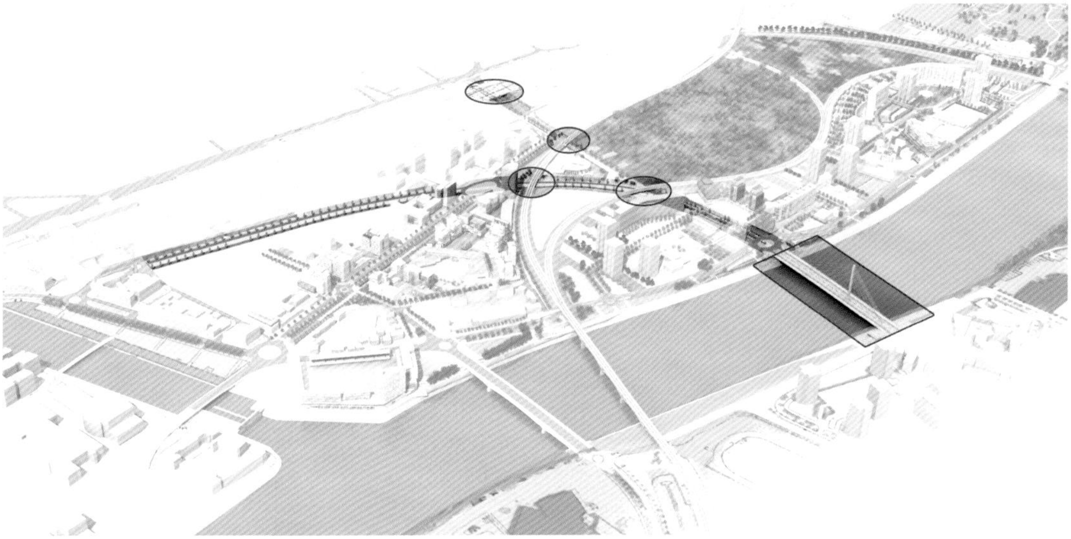

将街区融入城市中

打破马拉科夫基地的闭塞状态意味着让它重新和城市连结、便于穿越，同时也是让街区的生活更舒适，并改变它的形象。首先必须克服的是地形的和铁路的限制，因此方案整治了三座桥（两座新桥，一座拓宽）让人能够轻易地穿越铁道。此外，穿越卢瓦尔河的一座新桥和一条大道将马拉科夫街区和火车站及南特岛连接了起来。这条轴线改变了南特人的习惯，完全打破了过去居民遭到隔离的情况，并且改善了大众运输的服务。购物中心是项目的最后一块拼图，它必须能够汇集新、旧居民。

距离的缩短使马拉科夫得以融入城市的整体发展。与历史中心和卢瓦尔河的连结、朝向它们敞开的视野，都有助于拉近街区和城市的距离，同样地，明显而共享的公共空间也能够将城市的新旧空间结合起来（马拉科夫旧区、大群体社会住宅区、新建筑体）。

佩古谢区的建造改变了火车站后侧街区的状况，因而保证了和城市其他地方的连接。作为街区主干道的毕加索林荫道全面改变了街区的形象，并且吸引了投资商，也创造了生活的质量。

ANCHORING THE NEIGHBOURHOOD IN THE CITY

Opening up the site means both linking Malakoff to the city, being able to cross it, but also making daily life more comfortable and changing its image. It means first of all overcoming the geographical constraints and those linked to the railways. Three bridges (two new and one enlarged) make crossing the tracks easy. A new bridge over the Loire and an avenue connect the neighbourhood to the train station and the Île de Nantes (regeneration area on an island on the Loire). This axis changed the habits of people from Nantes, putting a firm end to the isolation of residents and improving the public transport service. A shopping centre, the last piece of the puzzle, should bring new and old residents together.

Shortening these distances has made Malakoff part of the development of the city. Connections and views to the historic centre and towards the Loire are part of this coming together, as well as shared and navigable public spaces, joining the old and the new (old Malakoff, the housing estate, new constructions).

Building the Pré Gauchet ensures the link with the rest of the city by changing the situation at the back of the station. The spine of the neighborhood, the Picasso parkway, radically changes the image of the area, attracting attention and creating a quality of life.

上图：位于铁路下方的新地下道和跨越卢瓦尔河的新桥
右图：位于马拉科夫街区东边、缓缓下降到卢瓦尔河岸的散步道，以及新建的塔巴里桥（建筑师马克·巴拉尼）的周边环境整治

Above: New railway underpasses and a bridge over the Loire
Opposite page: A walk in the east of the Malakoff neighbourhood leading down to the Loire, and facilities near the new Tabarly bridge (Architect Marc Barani)

"新"马拉科夫街区 · The "new" Malakoff

35

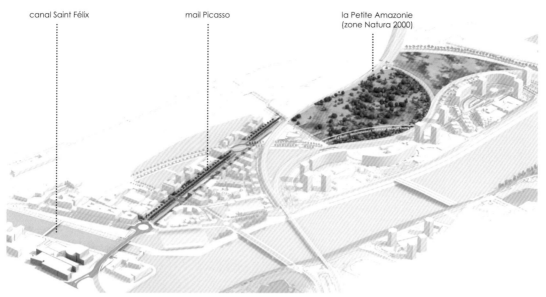

canal Saint Félix · mail Picasso · la Petite Amazonie (zone Natura 2000)

毕加索林荫道　　The Picasso parkway

39

地面的城市、行人的城市

城市的设计是从思考行人的视线移转和身体移动开始的。扩大行人对外界环境的感知，对城市生活的乐趣而言是相当重要的，必须透过建筑物和路径的美感来达成：如何从一个到另一个空间，设计是必须关注所有能够赋予空间流畅性、通透感的微妙细节。

对公共空间的关照同样延伸至地面楼层的通透性处理，以及活动和花园的存在。街巷工作室非常留意"小东西"，留心过渡性场所，这些不需要大费周章地改造但必须注意细节的空间。

毕加索林荫道被设计成一个渗透性很强的空间，表现在对于林荫道两侧建筑物以及横向越穿道路的处理上。建筑立面具有一定的视觉穿透性，小巷道则打开人们的视线。这种疏通开放的方式拉近了佩古谢区和卢瓦尔河之间以及马拉科夫高楼和新办公楼之间的关系。街巷工作室强调采用两种将马拉科夫与城市连接的方式：一则通过地面，创造一系列共同的城市符号（街道、人行道、地址、照明、小广场、花园、建筑前庭……），一则通过天空，从住宅大楼中寻找灵感。

THE CITY OF THE GROUND AND OF THE PEDESTRIAN

The starting point for designing the city is the pedestrian, from the movement of his (or her) eyes to the shifting of his body. Expanding perception is vital to the pleasure he derives from experiencing the city, thanks to the beauty of its architecture and his route: how to move from one area to another, with all these subtleties that give fluidity, porosity.

Care and attention with public space extends to the transparency of the ground floor, to the presence of activities and gardens. Atelier Ruelle applies this vigilance to the "little things" that do not require large changes but attention to detail, rather than large swathes.

The Picasso parkway has been designed as a porous space, with its banks and its crossings. Buildings must let people see through them, while alleyways open up views. This alternative method of opening up the area brings the Pré Gauchet closer to the Loire, and the Malakoff towers to the new office buildings. Atelier Ruelle proposes two ways of connecting Malakoff to the city: on the ground, by sharing the same urban codes (streets, sidewalks, addresses, lighting, squares, gardens…) and at sky level, drawing inspiration from the residential tower blocks.

1D Axo 街坊（建筑师Giboire-AIA）和2C1 Kanopé街坊（建筑师 AAUPC和ALDO）

Block 1D Axo (Giboire - AIA architects) and Block 2C1 Kanopé (AAUPC and ALDO architects)

法国 雷恩 / 自1999年起
permanence and consistency
持 久 性 与 合 理 性

多年来，街巷工作室和地方机构的对话已然建立，并且造就了一种强烈的都市文化。这个共通文化借助所有的参与者和高效率的规划工具而得以存在，为日后都市和城乡区域所选中的个别项目设计师们提供了一个重视持久性的工作框架。

于是，街巷工作室于1997年至2007年间为雷恩市的"肯尼迪楼板平台步行区"进行研究，并且完成了其"修复"工程。同时，街巷工作室也研究了该市公共空间的规划纲要（城市设计的一部分），在1999年至2001年间分成三个阶段进行：建立一份公共空间的策略性清单、建立指导蓝图（地标性场所和有关交通运输空间的主要规划原则）、在几个示范性基地上实际进行设计。与此同时，以水与河岸为主题的"蓝色计划"也重新展开。

Over the years, a dialogue has been established with a local authority that has built a strong urban culture. This common culture, supported by all stakeholders and by effective planning tools, provides a framework for contractors with whom the agglomeration and the city choose to work, with the intention of sustainability.

Thus, Atelier Ruelle conducted studies and directed the "repair" of the Kennedy Dalle, a concrete platform shopping area in the Rennes suburb of Villejean, from 1997 to 2007. In parallel, the group has studied Rennes' Planning Framework for Public Spaces (part of the city's urban project), set out between 1999 and 2001. This involves three phases: drawing up a strategic inventory of public spaces, drawing a main plan (sites of reference and guiding principles for public space related to circulation), and showcasing exemplary sites. At the same time, the Blue Plan has adapted this process for waterways and the banks of the river.

楼板平台的局部拆除以便通往温斯顿·丘吉尔大道，平台上公共空间的重新整治有利于连接地铁与商业空间，此外方案还进行了住宅户外空间的改造和停车场的重新组织

Partial demolition of the concrete platform to create a link to Avenue Winston Churchill, creation of a new public space in connection with the metro and the redesign of the commercial offer, reclassification of residential areas and parking restructuring.

肯尼迪楼板平台步行区
——维勒让街区的新中心

自从地铁通达此地后,维勒让大群体社会住宅区得以方便地对外联系,并和邻近的大学区拉近了距离。肯尼迪楼板平台步行区是更新计划中最重要的地点。商店从楼板平台的中心迁至一条通往雷恩市的新建道路,为新路带来生气;完全重新整治过的公共空间为新的设施(影音图书馆、社会服务中心)带来便利,并且透过私人花园而通往住宅;楼板平台下的停车场及其入口也经过重新设计。这个项目用去了不少时间来重新组织整体的机能。

Kennedy Dalle, the new centre for the Villejean neighbourhood

The extension of the Metro has opened up the Villejean housing scheme and brought it closer to the nearby university. The Kennedy Dalle is the centrepiece of this regeneration. Local businesses leave the centre of the concrete platform, instead bringing life to a new street with a real connection to the city. Completely reorganised public spaces benefit from new facilities (media library, community centre) and are accessible to residential property through private gardens. Underground parking has also been remodelled. The reorganization of all these aspects took time.

肯尼迪楼板平台步行区，位于温斯顿·丘吉尔大道的入口，整治前后面貌

The Kennedy Dalle, access from the Avenue Winston Churchill, before and after

街区主要空间——圣女贞德广场

公共空间的总体规划纲要变成城市发展的策略性文件，通过项目的落实而呈现出空间形态。以圣女贞德广场为例，新的空间几何强化出地面的不同高度，并扩大了教堂的前庭广场。方案并且限制了车速，保障了行人的舒适与安全。

Major areas of the district: Jeanne d'Arc Square

The public spaces master plan, which has become a strategic document, has taken shape in stages. For example, on the Jeanne d'Arc square, a new geometry emphasizes differences in height and the square in front of the church has been extended. Speed limits have been imposed for to improve the area for pedestrians.

从研究到方案——乔治·贝尔纳诺斯广场

乔治·贝尔纳诺斯广场是街区的另一个"主要空间"：平日作为停车场的长形空地可以转化为街区的节庆广场。广场东边的棕榈树公园衔接着莫合帕斯街区。

From study to project: Georges Bernanos Square

Place Georges Bernanos is another "important space" in the suburbs: a long esplanade usually used for parking transformed into a festive neighbourhood square. In the east, a palm garden connects it to the Maurepas neighbourhood.

"蓝色计划"在圣马丁草原成形

"蓝色计划"建议从最自然的河岸风光逐渐过渡到最城市化的河岸景观。2012年街巷工作室参加圣马丁草原的设计竞赛，所提交方案的创作灵感便来自"蓝色计划"的原则。圣马丁草原区邻近雷恩市旧城中心，是容易泛洪的河套区，极少被开发使用。扩大此地洪泛平原面积的需要促使了一个想法的诞生：在此创造一个呈现自然风貌的都市公园，以尊重水文和历史的印记：花园、工业及生态环境。

The Blue Plan takes shape in the Saint-Martin Meadows

The Blue Plan sets up a gradual progression of the banks of the river, from the most natural to the most urban. This was the principal behind the response to the Saint-Martin Meadows competition, which called for the transformation of a meandering body of water, ill-adapted to the desired uses and liable to flooding at the edge of the historic centre of Rennes. The need to expand the flood plain gave rise to the idea of creating a natural urban park, which respects the path of the waterway and its history – gardens, industry, habitat.

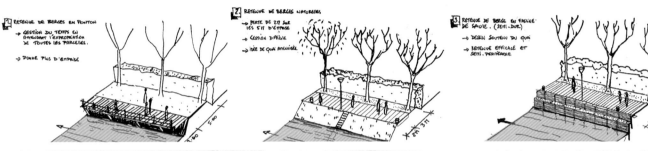

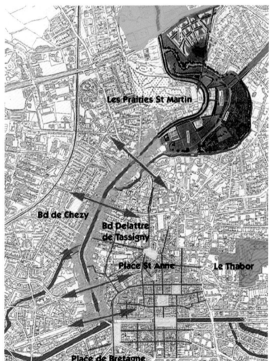

上图：摘取自"蓝色计划"的图面资料，1999
右图：圣马丁草原的整治方案，2012 年设计竞赛

Above: Extract from the Blue Plan ,1999
Right: Proposal for the development of the Saint-Martin Meadows, 2012 competition

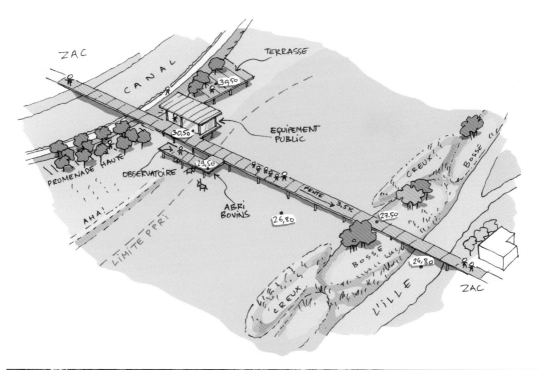

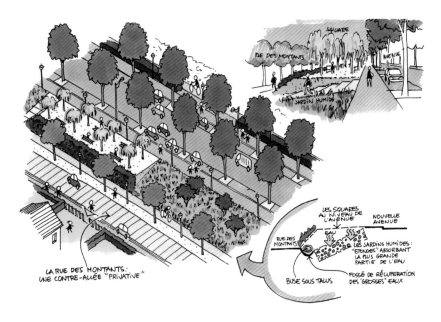

法国 圣迪吉耶 / 2001-2013

urban strategy
城市策略

与其提出一个规整的现成方案，街巷工作室选择了提出一个方法和策略，以辨识出项目中能够产生干杠效应的有利层面，并以微妙的手法处理最棘手的领域。圣迪吉耶的城市大计划从2000年就已经展开，面对着去工业化的转型过程及居民迁徙郊区的情况，这个计划出自市长弗朗索瓦·考努–让提耶的意愿。第一届任期内，他先致力于重造市中心的吸引力，接而着手重整"绿树林"街区，1万名市民（占一半的城市人口）就住在这个社会住宅区里。

街巷工作室首先把重点放在消除4号国道带来的障碍——这个在实体与心理上都把社会住宅区与都市分隔开来的屏障，让所有的交叉路口的道路在同一个高度上交汇，将其还原为供人交流往来与通行的地面。接下来，工作室试图寻回原有的自然景观，这片介于森林和马恩河谷间的丘陵地带，随着过往一栋栋建筑物的建造而逐渐丧失了优美景致。一些旧建筑物的拆除与新公共空间系统的建立使得地形地貌得以重现出来。接着需要的是耐心：由于房产市场低迷，重建的数量仍然不足以让房屋提供多元化的选择。况且还有商业中心这个难题要处理。

在等待期间，圣迪吉耶于2012年推出了市中心与马恩河岸整治的城市设计项目，方案交给来自巴塞隆纳的卡梅·皮诺斯团队负责。这个新的城市理想之所以能够有所发展，全得归功于经过了10年终究能够共享的方案文化。

Rather than a rigid pre-prepared project, Atelier Ruelle proposes an approach and a strategy that identifies leverage and tackles the most difficult issues sensitively. On the initiative of Mayor François Cornut-Gentille, the Saint-Dizier GPV (Grand Projet de la Ville, or grand city project) regeneration project was launched in 2000 to deal with de-industrialization and the exodus of residents to the suburbs. In his first term of office, the mayor improved the attractiveness of the city centre, and then took on Vert Bois, a social housing neighbourhood for 10,000 people, half the local population.

Atelier Ruelle's work benefited first of all from the removal of the RN4 trunk road, a physical and mental barrier that separates the area from most of the city. This went together with levelling all intersections, and returning the ground to a place where people can walk and exchange. Next, the team looked for ways to restore contact with the original landscape – hills between forest and the valley of the Marne, whose views had been lost through successive construction projects. Demolition and a new framework of public spaces reveal the topography. Patience was required after this: a lack of demand on the housing market meant there were not enough new builds to diversify the supply of housing. And the difficult question of the shopping centre remains to be addressed.

Meanwhile, Saint-Dizier launched an urban project in 2012 focused on the centre and the recovery of the banks of the Marne, which was awarded to Barcelona architect Carme Pinós's team. A new ambition for the city was made possible thanks to a project culture that, after ten years, has become a shared approach.

SUPPRIMER LE "MUR" DE LA RN4!

- mise à niveau des carrefours :
  des espaces publics dans la ville
  Upgrading crossroads: public spaces in the city
- création de traversées complémentaires
  Creation of additional crossings

法国 奥利 / 自2002年起

the suburbs also have a history
郊区也拥有历史

街巷工作室因为反对"郊区没有历史"的想法而产生了"沉积"的工作方式，将他们的规划设计工作融入了由市长加斯东·维扬及城市规划师让·得罗什共同建立的方案文化与默契之中：30年来所进行的城市更新、艰难的社会住宅区整治行动，以及一条将老村落、大群体社会住宅区和塞纳河连接起来的城市轴线的建立。

2002年，赢得项目规划研究竞赛之后，街巷工作室便介入了"飞行员"街区和"牧师之石"街区的整治，其工作包括为街区的都市设计项目进行构思与执行、对几个社会住宅社区进行改造、替某些公共空间进行设计并交由市政单位完成施工……2009年以来，工作室与市政府和开发商瓦罗菲斯、艾克斯庞西耶勒合作，针对几个协议开发区之间的缝隙与剩余空间做了一个全面性的评估与省思。从这个新尺度出发，该项目不再拘泥于对大群体社会住宅区主题的研究，而进入人文关怀的层面与措施，因此得以远离所有贬低的诠释而和郊区的精神重新产生联结。

A sedimentation approach has countered the idea that the suburbs do not have a history. Atelier Ruelle adheres to an enduring project culture and complicity established by a mayor Gaston Viens and his planner, Jean Deroche: thirty years of urban renewal, of difficult interventions in social housing areas, the construction of an axis, the Voie des Saules (or Path of Willows), which connects the original village to the huge social housing district and the Seine.

After winning a competition in 2002, Atelier Ruelle has been working on the Aviateurs and Pierre-au-Prêtre neighbourhoods. This work includes designing and monitoring the urban project, designing the exterior space of several social housing units as well as public spaces in collaboration with the City services. Since 2009, the group has worked with the local authority and the developers Valophis Habitat & Expansiel towards a global appraisal of the cracks between the ZACs (comprehensive development zones) and residual spaces. On this new scale the project could leave the narrow thematic of the social housing scheme behind and adopt a humanistic approach that revives the spirit of the suburbs, far from any pejorative interpretation.

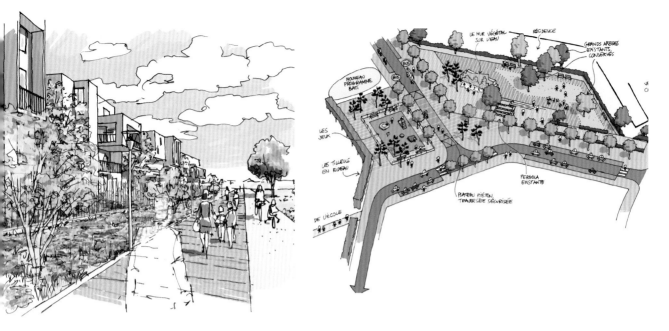

pleasures
享 受 乐 趣

大海、公园、河流、自然的存在……有时，基地准备好要打出空间牌，展现场所的丰富优势。有时，这个关系变得比较微细、脆弱，尽管如此，重寻都市的乐趣仍是可能的。重新投注于对城市之爱的建立，激发人们想要在此定居、留驻的欲望，此目标变得刻不容缓，行动也成为关键所在。

城市规划的工作不在于卖弄伎俩或企图为都市进行推广宣传，而是确认空间的用途，包括日常生活所需和有利身心舒畅的空间，还有节庆用途、特殊时刻、消遣娱乐以及令人愉悦的空间。

想让人喜爱一个场所，便得展现它的特色，显示它之所以与众不同的地方，这是不言而喻的道理，却时而被人所遗忘，需要被再度点明，就如同人们进行自我评估的工作一样。

The sea, a park, a river, the presence of nature... Sometimes a site naturally displays qualities of space, a generosity of place. Sometimes this link is more tenuous and fragile, but it's possible for a town to reconnect with this pleasure. It's important, because there is an urgent need to reinvest in the love for a city, generating a desire to live there and stay there.

Far from a gimmick or a city branding campaign, the work of urban planning justifies certain uses of spaces: those of everyday life and well-being but also festive customs, entertainment and contributing to feelings of happiness.

To make a place loved is to bring out its character, to reveal what makes it remarkable, an obvious fact that sometimes is forgotten and needs to be made clear. It is like working on one's own self-esteem.

法国 维勒雷洛斯 / 1999-2013
simple pleasures
单 纯 的 乐 趣

离艾特尔塔不远的维勒雷洛斯这个滨海小站曾是雨果的栖息地，在第二次世界大战期间饱受炮火摧残，战后经过重建，将维勒河整治为渠道，并建造了一个停车场覆盖其上。在海岸边和陡峭的悬崖脚下，维勒河这条法国最短的河流的出海口在此形成一个洼地，这个美丽的基地已然寻回过去的魅力，大众海滩在此诞生，河面也重显波光粼粼的景致。

一座大型的木质甲板，令人联想到远洋游轮，为此地单纯的乐趣增添了光彩：线条圆滑而舒适的长椅、往海面延伸的观景平台、为孩童建造的游戏空间与浅水游泳池……都为日常生活带来了美感。长条形的座椅环绕四周，呈现出整体合一的空间感。单纯，便是信任空间本身，无须再画蛇添足。

Close to Etretat, the small resort of Veules-les-Roses is where Victor Hugo found refuge. It suffered from bomb damage during the Second World War and then from the post-war reconstruction, which channelled the Veules and covered it with a parking lot. On the waterfront at the foot of steep cliffs, the mouth of the Veules, the smallest river in France, has carved out a sheltered space: this beautiful site has been given back its appeal, it now has a popular beach and sunlight once again glints on the river.

A large wooden deck, evocative of an ocean liner, increases the simple pleasures of the place, enhancing everyday life. There are rounded and comfortable benches, a viewpoint on a pier extending into the sea, playgrounds and a paddling pool for children. Long benches surround these areas and give unity to the whole. Simplicity is to trust the space itself, without adding to it.

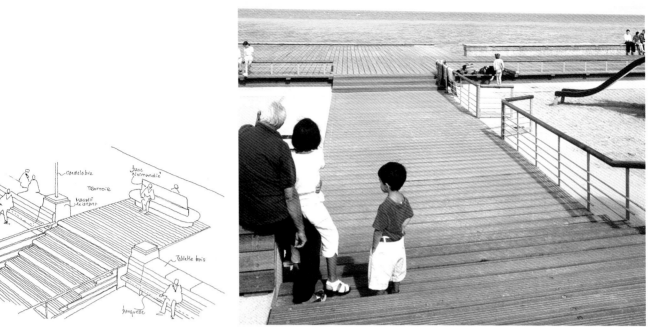

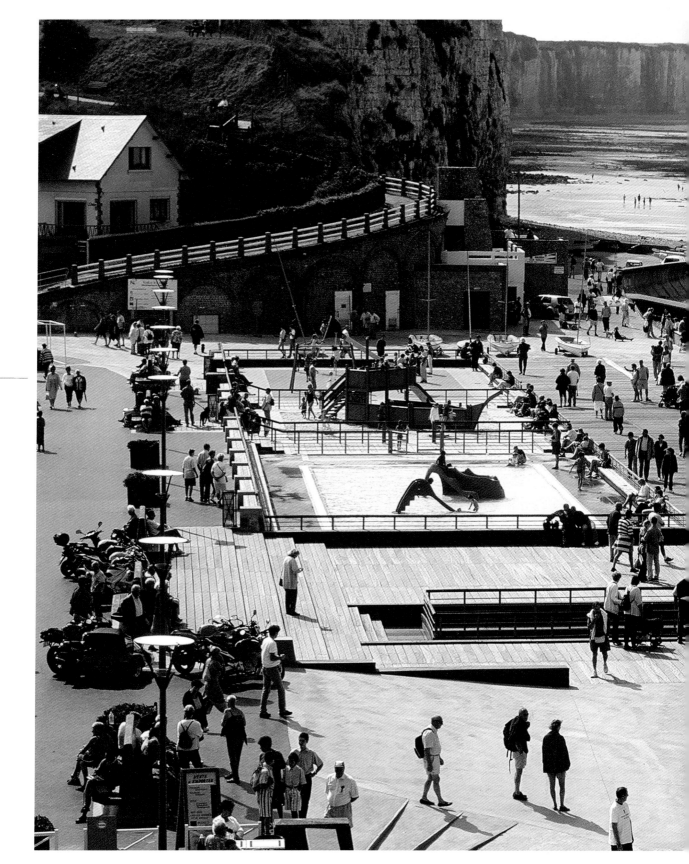

解放维勒河

新阶段的规划（2012-2013，进行中）重现了过去改为水渠而且被停车场覆盖的维勒河风貌。停车场的拆除使得河流两侧坡岸形成了淡水花园，而位于水渠入海口的卵石海滩边上则设置了鱼梯。

Liberating the Veules

The new phase (ongoing 2012-2013) uncovers the Veules, which had been channelled and hidden under a car park. Breaking up the car park gave way to a freshwater garden along the river. A fish ladder has been fitted at the mouth of the river on the pebble beach.

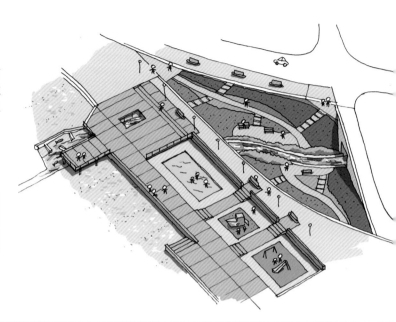

65

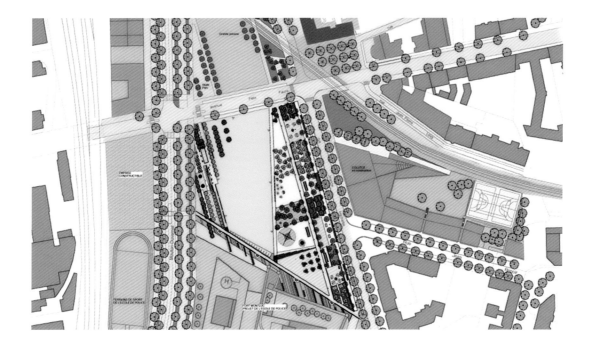

法国 里昂 / 2005

fluidity and intimacy
流 畅 与 隐 秘

如何能够同时保证玩耍自由和安全保护，表达流畅又隐秘的感觉？这个公园离帕尔迪约火车站不远，是蒙吕克城堡脚下一个宽阔广场经过改造的结果：它提供了一个自由开放的空间，同时又不受到周围街道与铁路的干扰。其流畅性来自对广场边界的构思工作，这是先经设计绘制再逐步解构的结果。人行道变成绿树成荫的花园，花园通往游戏区，最后抵达一片草地。

广大的草坪上阳光闪烁，吸引了绿荫步道上行人的注意。行人从一个世界进入另一个世界，走过典雅静谧的花园，穿越整治为游戏场的林中空地，逐渐下降而抵达草坪。这些林中空地向着宽阔的草坪倾斜，通过人体的刻意弯倾则更加产生一种朝向草坪前进的动力。

How to express both freedom to play and protection, fluidity and the feeling of intimacy? This park transforms an esplanade at the foot of Fort Montluc, near the Part Dieu station. It provides a free and open space, while protecting from the surrounding streets and railways. The fluidity is the result of working on the boundaries, which were first drawn and then gradually deconstructed. A pavement becomes a shaded garden, the garden opens onto playgrounds and finally on a meadow.

The pedestrian moves from shady pavements, attracted by the light of the large lawn. He (or she) passes from one world to another, through a formal and tranquil garden to a converted clearing, until gradually he reaches the sloping meadows. The clearings give way to a large lawn, playing on the inclination of the body so that this in turn inclines towards the meadow.

属于街区的空间

这个向街区敞开的公园是可以自由进出、让人放松身心的空间，它将这块离市中心不远但临近铁路的荒地转换成有用的场所。它位于一所中学对面，提供了许多游戏设施和一片宽广的草坪，人们可在草坪上玩球、做日光浴、进行野餐。

A neighbourhood space

The park, a public recreation area open to the neighbourhood, forms part of the reclaiming of a strip of wasteland near the centre, by the railway lines. Facing a school, it offers a children's playground and a large open lawn where people can play ball, sunbathe or organise picnics.

宽广人行道转化为花园

从人行道到草地,从绿荫到阳光,从隐秘到开放。

行人们从植树人行道走入交错分布的绿篱,而缓缓进到了花园。这些位于边缘地带的连续性绿篱首先塑造了一个隐秘的绿荫空间;接着林木转而稀疏,呈现在眼前的是整治过的实用性空间,摆设了桌子和长凳;这些林中空地最后对外敞开,通过游戏区和喷水池而面向着广大的草坪。

这个空间运用地面的高低配置和错位关系提供了多样化的用途与行进空间。

A large pavement becomes a garden

From pavement to meadow; from shade to light; from privacy to openness.

The tree-lined pavement becomes a garden into which the pedestrian slips, the shift marked by staggered hedges. This border of successive hedges defines a space that begins intimate and shaded. This shade then thins out, giving way to spaces equipped with tables and benches. Finally, these clearings open out, through children's playgrounds and fountains, to the great lawn.

The space plays on different levels and staggered ground to provide a wide variety of uses and movements.

从左到右:公园边缘的人行道 / 树荫下的花园 / 倾斜的林中空地 / 大草坪

From left to right: The pavement as a fringe / A shaded garden / The clearing on an incline / The great lawn

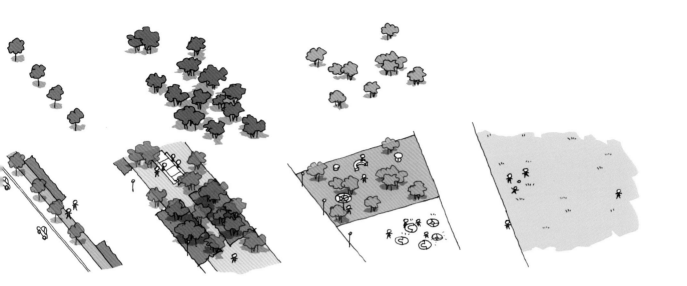

主题的变奏

为这个公园设计的小品结合了实用与趣味的需求，一些基本元素相互搭配而形成各种不同的组合：长条椅搭配树木或独立存在，也可配合路灯、桌子或喷水池……

Variations on a Theme

The furniture custom-designed for this park allows for practical requirements whilst remaining playful. It combines basic elements – benches with or without trees – with lighting, a table or a fountain.

法国 蓬图瓦兹& 圣鸟昂洛莫纳 / 2004-2011

the meaning of the site
基 地 的 意 义

在这个位于城墙脚下的瓦兹河历史性堤岸上，一个全新的码头花园为废弃的旧港河岸的改造开启了崭新的一页。花园融入了矿物岩层的景观中——码头、石墙、水平延伸的板岩屋顶、城墙，展示出河流及周边地区的丰富关系。项目的功能计划包括了几个公共空间的设计以及一个旅游服务中心与港口管理办公室，为蓬图瓦兹和河对岸的圣鸟昂洛莫纳在未来可能建造的天桥做准备。

起伏不平的地表间，一座历史悠久的桥梁串接了河流上游较高的河岸，下游较低的河岸上方则有铁路桥通过，旅游服务中心就设置在低地上。它的屋顶缓缓上升，成为一个观景平台，未来的天桥可以和这个屋顶平台连接。一条散步道低调轻巧地滑向建筑物的屋顶，此屋顶成为公共空间的一部分，却不会阻断散步道与瓦兹河的关系。

这个项目是最能代表街巷工作室处理手法的方案，在此，建筑尺度和公共空间相互交错连接。

On this historic bank of the Oise, at the bottom of the ramparts, a new wharf-garden begins the reclaiming of the riverbank from the old abandoned port. The mineral strata of the landscape – banks, stone walls, horizontal slate roofs, ramparts – are echoed in the park, which brings out the rich relationships between the river and the surrounding area. The programme involves the creation of public areas and a Tourist Office-cum-Harbour Master's Office. It paves the way for the possible arrival of a footbridge between Pontoise and Saint-Ouen l'Aumône, on the other side.

Between the folds of the gradient, between the ground further upstream connected to the historic bridge and the land downstream that passes under the railway bridge, stands the Tourist Office on lower ground. It rises gradually, its roof terrace serving as a viewing terrace, on which the future bridge will hang. The footpath slides discreetly over the top of the building, which forms part of the public space without blocking the link to the Oise.

It is a flagship project emblematic of the Atelier Ruelle approach, where architectural dimension and public space are intertwined.

建筑物变成公共空间

旅游服务中心处于两层地面（码头堤岸和未来的天桥）之间，连接了坚实的石头地面和轻盈悬空的木质天桥。地面的连续性和室内外的透明关系保证了建筑物和公共空间之间的通透性。

室内的遮阳板时而过滤着窗外的市景，时而成为展览所需的配备。

Architecture becomes public space

The Tourist Office covers two floors, at the level of the wharf and that of the future bridge. It sits between firm stone ground and the footbridge made of wood, which is light and airy. The porosity between the building and the public space emphasises the continuity of the ground, and the transparency between interior and exterior.

Inside, sun-screen shutters filter views of the city or become exhibition supports.

法国 巴黎 / 2005-2007

come into Paris
进 入 巴 黎

街巷工作室非常喜爱巴黎的公共空间，得以在此项目展现它们的优点：就整个都市的尺度而言，它们的规律性在法国是独一无二的，这也赋予巴黎的公共空间绝佳的辨识度；它们丰富多样而具有弹性，可以在细节上产生众多变化，例如在一个多地面层次的基地上，可以出现多样化的植被形式与花坛设置方式。

文森门大道是文森林荫道的一部分，往来的汽车（繁忙的交通、加油站）使它一度丧失了林荫道的特质，如今通过整治则得以恢复其作为历史轴线的崇高地位。大道两旁社会住宅区的条状建筑也因此重拾宁静，这全归功于公共空间的舒适品质：在非常宽阔的人行道上设置的花园、为行人与单车设计的安全路径。后者在此尤其重要，因为在地铁、公车、轻轨电车、前往文森森林的散步道……之外，此地的软性交通（行人、单车、滑轮等）非常频繁。

Lovers of Parisian public space, the team at Atelier Ruelle had an opportunity here to demonstrate its worth. Its consistency, unique in France at the level of an entire city, gives it a remarkable legibility. At the same time, its richness and fluidity allows for variety at the level of detail, such as the diversity of tree-planting, or how a flowerbed comes out of the ground, in a fold of land.

The Porte de Vincennes Avenue, a section of the Cours de Vincennes, had been sacrificed to the car (heavy traffic, service stations), losing much of its appeal. Here, it regains the majesty that comes from being a historical axis. The high-rise housing estates that surround it have regained a serenity thanks to the amenity of public spaces: green spaces on a wide sidewalk, protected pathways for pedestrians and cyclists – particularly important here where pedestrian traffic is intense, between the metro, bus, tram and the walk to the Bois de Vincennes.

宽阔舒适的散步道

人行道是专为行人或单车骑士保留的空间，经过大幅拓宽、植树以后，变成了一道线形花园，中间错落几个小广场，边上则是住宅区花园，以此过滤大道和住房之间的关系。花园的节奏柔化了规划线条的对称性，一如花园的多样氛围为视觉的连续性带来丰富的变化。大道两边一栋栋条状建筑和其山墙时而交替，大道上的花园组织则与此相互呼应。

为了节约资源和简化手法，方案保留了原来的树木，新植树木时也不干扰既存的设备系统，并且极力回收材料和增加可渗透雨水的地面。

A spacious and comfortable walk

The sidewalk, generously expanded with planted areas, is reserved for pedestrians and cyclists. It becomes a linear garden, punctuated by public squares and bordered by residential gardens that filter the relationship between the avenue and housing. The symmetry of the design is softened by the rhythmic quality of the gardens as visual continuity is enriched by the variety of environments, composed as an echo of the architecture of the high-rise blocks that line the avenue, alternating with their gables.

The desire to save resources and means meant we sought to keep the existing trees, planting without disturbing the utility networks, recycling materials and increasing permeable surfaces.

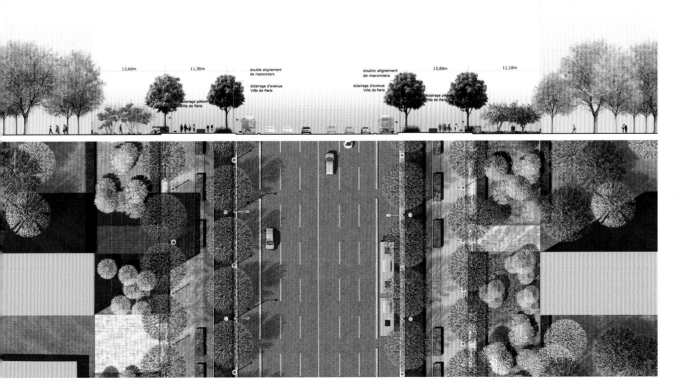

重新整治的立交平台

环状的立交平台原本被汽车道路从中穿过,如今它一如文森门大道,也经过重新配置和整治,为行人与单车路径提供方便:其人行道拓宽了,环形区的中央则禁止汽车通行。汽车行驶空间被减少为原来的一半,却不会引起塞车。

这个方案重新建立了介于巴黎、周边城镇和文森森林之间的行进连续性。

The interchange, too, has been reconfigured

Before, the interchange ring was crossed by a multilane highway. Like the Avenue, it has been reconfigured to allow for pedestrians and bicycle routes: wider sidewalks, the ring closed to traffic. The area devoted to cars has been halved, without causing traffic jams.

The continuity of routes between Paris, neighbouring municipalities and the Bois de Vincennes is restored.

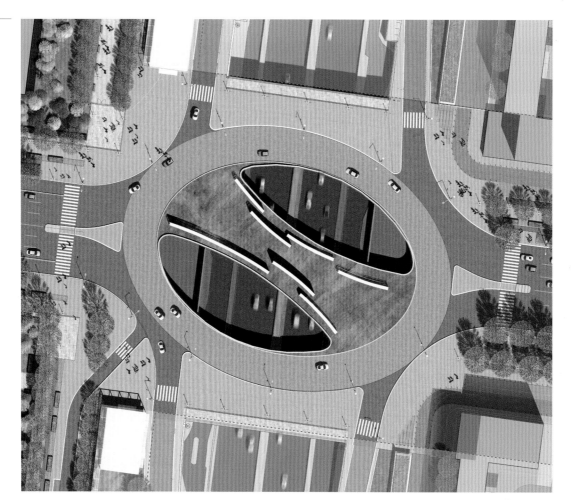

法国 第戎 / 2012

campus spirit
校 园 精 神

第戎的大学校园位于市区内，自1950年代起，大学街区便环绕着伊拉斯谟广场逐渐建造成形，广人的草地由一群不谐调的建筑物包围着，有的面对着它，有的则背对着它。原先规划设计的连贯性已然丧失，空旷的广场空间被当作停车场使用。想要在此重寻城市生活的乐趣，便必须将一系列的剩余空间改造成线性公园，将汽车排除在外，享受这个几千名学子与研究人员活动往来的街区氛围。

不论在个人生活或团体生活当中，这里的空间都允许人们以自由轻松的心情来表达自己，这便是校园精神显现的方式。宽广而大方，既开放也隐秘，每个人以各自认为合适的方式来自由享用这个空间，不论是走遍或穿越校园，在其中独处或与人会面。

In Dijon, the campus is in the city. Since the 1950s, the university district has been built gradually around the Erasmus esplanade, a large meadow surrounded by assorted buildings, some of which are turned towards it, others turning away. The coherence of the original plan is lost, the vast empty space of the plaza has been invaded by the car. To regain the pleasure of the urban means transforming a series of residual spaces into a linear park, removing cars and making the most of the energy of a neighbourhood frequented by thousands of students and researchers.

The spirit of the campus is brought forth by the way the space allows individuals as well as groups to express themselves freely and in a spirit of light-heartedness. Ample and generous, both open and intimate, it offers everyone the freedom to enjoy it as they see fit – as a means of getting from A to B, a place to be alone, or to meet friends.

不求整齐规律，但求通透流畅

经过2004年的规划研究竞赛之后，随着阿尔弗雷德·彼得设计的轻轨电车延伸至大学校园基地，街巷工作室得以针对其中央广场进行整治。

这是一条向南面伸展的线形空间，排除整齐规律的布局，这个空间并非以建筑立面而是以边缘处理来作为界线。建筑物本身也有通透性和活动力，其内部的生活不仅在在研究或教学单位里面发展，也扩张到户外的公共空间。

Not rigid, but permeable

After being selected in a public competition in 2004, Atelier Ruelle was able to create the esplanade following the extension of the tramway to the university campus, a line designed by Alfred Peter.

Stretching for a long way towards the south, there is no rigidity in the layout: the space is not bounded by walls but by edges. The spaces between buildings are permeable, allowing the internal lives that develop within research or teaching units to overflow into the collective space.

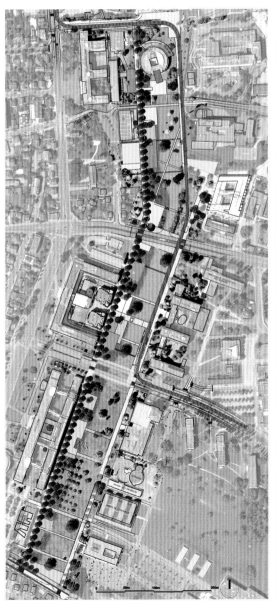

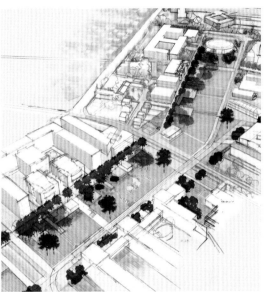

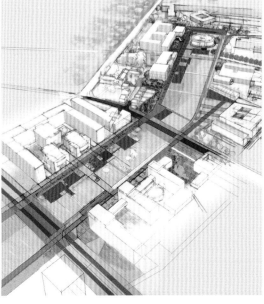

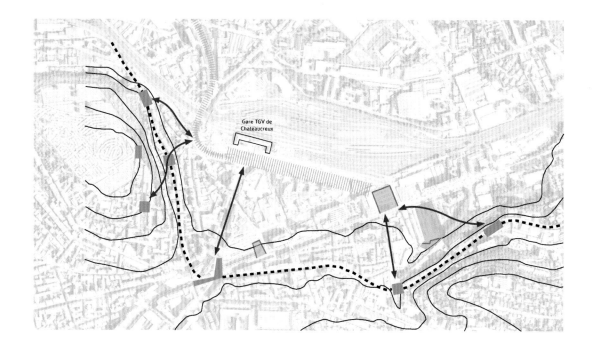

法国 圣艾蒂安 / 2008-2017

rediscovering geography
重寻地理面貌

重新探索都市、市容景观、地面、斜坡…… 圣艾蒂安所拥有的丘陵地潜力被米歇尔·寇拉儒比喻为城市的宝藏："透过城市和自然景观所维持的紧密关系，城市本身也变成了景观。"凸显随着等高线而设置的街道路径、保护远眺的视野、框取面向自然大地的景致，这些都有助于城市自身景观的呈现。

街巷工作室便是如此构思沙陀克火车站街区的，使其与广阔的景观相呼应，这个景观不论是从山下低凹处或从周围山峰都能观赏得到。办公楼和住宅的兴建项目将有助于塑造高低起伏的景观形态，不同的观景台也将凸显山丘郁郁葱葱、丰富而多样的景致。

这个着重基地的工作不仅适用于大尺度的空间，也能运用于小尺度的场所，借此设计一个小广场，凸显不同的地面高度，建造交叉路口的花园，保护山坡，探索一条过道、一个街坊中心、一个庭院……

Rediscovering the city, its landscape, the soil, a slope... At Saint-Etienne, this is the strength of a hilly site that Michel Corajoud describes as an urban treasure, in this city that "by the intimate relationship that it has with its landscape, itself becomes landscape". A landscape that is seen by revealing the pathways of the streets that follow the contours of the land, preserving the distant horizons and framing the views of the natural horizon.

This is the way in which Atelier Ruelle conceives the station area, Châteaucreux, in accordance with the scale of the landscape, perceived as much from the hollow below as from the peaks that surround it. Projects for offices and homes will help sculpt the terrain, viewpoints will highlight the diversity of views to the wooded hills.

The work on the site plays to both large and small scales, to design a square, highlight a difference in height, make a garden at a crossroads, preserve a hillside, discover a passageway, the heart of a city block, a courtyard...

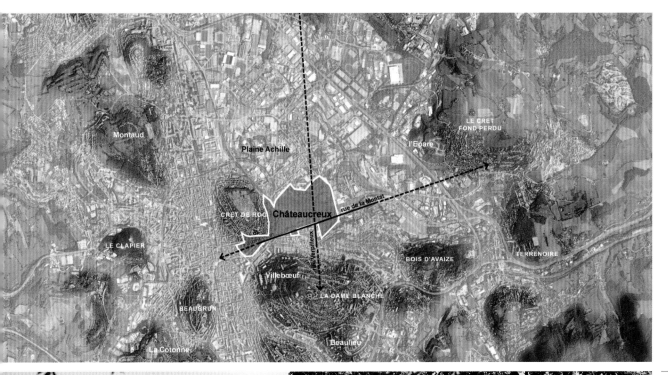

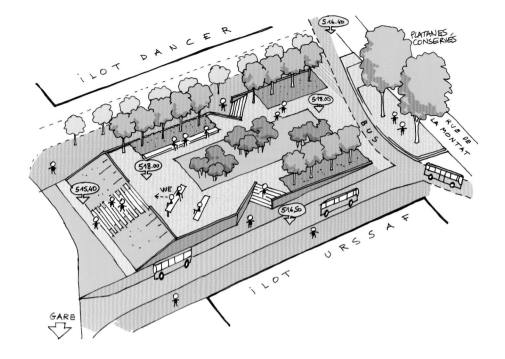

on our doorstep　　　　自 家 楼 下

法国 昂热，勒蒙，南特，圣丹尼，圣纳泽尔，热纳维利埃

paying attention to everyday spaces
关 注 日 常 生 活 空 间

都市是文明礼仪的象征。将这个文明的语汇拓展到每个街区，包括闲置废弃的空间，这便是规划公共空间这项工作的使命。这台"压路机"，这个必须到处伸展的共通语汇，要让所有社会大众都能使用，无论项目规模是大或小：一个花园、一个广场、一整个社区……

城市规划师的作为就如同"规范师"，他知道自己使用的是具有调和性的空间语汇，也清楚他所建立的参照效果，虽然偶尔会刻意偏离规则。公共空间能让人自由地与他人同处，将距离与分享结为一体。

街巷工作室为大型平民街区所做的规划设计便是基于这个公民的意愿。观察人们如何使用空间，倾听居民的想法却不给予直接的回答，因为有时必须说服居民接受某些只有他们实际经历过后才会赞许的选择。

The city is civility. To extend this civil language to all neighbourhoods, including places kept out of sight, is the mission of our work on public space. It's a steamroller, spreading this common language everywhere so all the public can use it. This is valid whatever the scale of the project: a garden, a place, an entire city...

The planner acts as a "normaliser", aware of the peacemaking vocabulary that he uses and the references it produces – sometimes he (or she) even uses it to deviate from the rules. Public space can give freedom to be an individual with others, to combine distance and sharing.

The interventions of Atelier Ruelle in large, working-class districts are based on the desires of citizens, observing how people use their space and listening to what people say, without answering directly because sometimes it is necessary to convince people of choices that will only be appreciated when they are lived.

大型平民街区的问题并非由于欠缺良好的规划意图，而是疏忽大意所致，以造成缺乏连贯性与长久的维护。要为其重新进行整治，意谓着恢复最初丰盛大方的美好意愿，同时搭配植被计划，找回迷失的方向或变得模糊的识别性：为空间带来"第二层"整治，以等待未来继续加诸于上的其他改造。这第二代整治工作着重于空间的使用与管理，空间的使用性可通过观察而获得，因为居民已经在此生活，而为了维持日常生活的品质，必须保持简易的管理和低成本的保养维护。

为了居民的舒适、安全，并使他们关注介于公众与私人领域之间的微妙场所，公共空间对群体生活带来极大的助益，也或许能够促进一种尊重他人的社会行为。

Large working-class neighbourhoods do not suffer from a lack of good intentions in the first place, but from carelessness, lack of consistency and care. To intervene means regaining the initial and generous goodwill by completing the vegetation, recovering lost directions or meanings that have become blurred: adding a "second coat" in preparation for future additional layers. This coming of age pays attention to how people use the space – something that can be observed as neighbourhoods are inhabited – and how it is managed, which should allow quality of life to be fairly easy to maintain on a daily basis at low cost.

Through their comfort, their security, the attention paid to the sensitive areas that establish the limits between the public and private domain, public spaces contribute to community life and can influence whether or not people behave respectfully towards each other.

关注建筑物周边空间的质量以及边界的处理，昂热的勾贝尔街区

Quality public space at the foot of buildings, designing borders carefully. Angers, Gaubert neighbourhood

地面处理

材料的质地，或光滑或略粗糙，或多或少被植被所占据，因而或多或少具有孔隙以便渗透或回收水分，这就是地面的处理，是感官接触的所在，行人应该能在这里体会丰富的感受，感觉到材料本身和它对光线的反应，并且看到其所行走的地面如何和建筑物建立关系。

对市政单位而言，地面处理为公共空间带来多种新的维护方式。

Designing the pavement

The texture of materials, more or less smooth, more or less colonised by vegetation, porous in order to filter and recover water: the ground is the place of contact where the pedestrian should feel a wealth of sensations, from the materials and their reaction to light to how the surface you walk on relates to the buildings.

For city services, careful work on the pavements brings new ways to maintain public spaces.

上图：昂热，玫瑰园街区
右图：勒蒙，格罗尼耶尔街区

Above: Angers, the Roseraie (Rose Garden) neighbourhood
Right: Le Mans, Glonnières neighbourhood

路径的流畅性、清晰度

公共空间的首要任务是使行走的路径流畅与清晰,让每个人都能轻易理解路径的安排组织。它们制造连续性,而非同质性,建立城市和街区之间的关系,创造与汽车隔离的行人路径的舒适感……

简约行事是必须的准则,如此才能以最少的成本获得最大的使用面积,同时预先考虑维护事宜。简约行事也是一张王牌,可以避免针对特定用途来塑造空间。

Fluidity, clarity of pathways

The first mission of public spaces is fluidity and navigability so that their organization is easily understood by all. They should be continuous rather than homogeneous, making links between neighbourhoods and the city that surrounds them, the comfort of pedestrian paths separated from cars.

Keeping it simple is necessary to get the most of out the space with minimum costs and anticipating maintenance. An asset that avoids specialized uses.

上图:勒蒙,格罗尼耶尔街区
右图:南特,马拉科夫街区

Above: Le Mans, Glonnières neighbourhood
Opposite page: Nantes-Malakoff

107

属于孩童的游戏场所

这些平民街区的小孩数量很多，必须在这里设置游戏设施、长凳、树木……一如所有人潮频繁的场所，这些空间需要的是简约大方的处理，才能使它们拥有个性，凸显特殊场所的性格，通过热闹空间和安静空间的轮流设置为街区的日常生活带来节奏。

Children's playgrounds

There are many children in these working-class neighbourhoods. They need playgrounds, benches, trees... Like all heavily used places, these areas require simple but generous treatment to give character, highlight specific sites and keep in harmony with daily life in the neighbourhood by alternating between lively places and quiet spaces.

上图：南特，马拉科夫街区的游戏空间
右图：勒蒙，格罗尼耶尔街区

Above: Nantes, Malakoff playgrounds
Opposite page: Le Mans, Glonnières neighbourhood

边界处理手法

为了不使安全设施的限制成为设计的困扰,最好的方法不是排拒它们而是与之共处,利用它们来设计空间的结构线条、延伸地面或植物的点缀。由此,隔离用的金属网塑造出的是格栅围栏的效果而不是封闭的隔墙,它们具有筛滤效果,使得周围环境相互渗透。

Playing with limits

So that security constraints are not severe, it is better to incorporate them rather than to fight against them: to use them to draw paths, extending the ground or plant beds. In this way, metallic threads create trellis fences rather than enclosures thanks to their filtering effect, which permeates with the surrounding space.

上图:南特,马拉科夫街区的游戏空间
右图:圣丹尼的儿童权利广场,以及南特马拉科夫街区的游戏空间

Above: Nantes, Malakoff playgrounds
Opposite page: Square of the Rights of the Child in Saint-Denis and Nantes, Malakoff playgrounds

111

安置汽车

为了避免汽车的存对行人步行的乐趣与方便造成干扰，必须将停车场设置在适当的地方，并使其融入新的都市景观之中。安排和整治停车空间对日常生活的舒适度和对街区的形象而言都是不可或缺的措施。

What to do with cars

So that the presence of cars does not get in the way of the pleasure and ease of use of pedestrian routes, car parks are needed in specific places, integrated into the new urban landscape. Organising and redefining car parking is an essential step for the comfort of daily life, just as important as the image of the neighbourhood.

南特，马拉科夫街区的停车场
Nantes, Malakoff carparks

在花园里生活

街巷工作室主张大量种植，因为除了为感官带来愉悦的感觉，植物也可以为建筑物和公共空间重新设定边界，为场所塑造不同的面貌与氛围，为人创造惊喜，并在既有植被的基础上结合新与旧……

提供给居民这些宽阔大方的空间，也便是重新找到大群体社会住宅社区的原始目标：尝试以开放的空间带来生活的质量。

Living in gardens

Plant abundantly: in addition to the sensory pleasures they provide, plants recreate the boundaries between buildings and public spaces, diverse places and environments, are full of surprises, and combine the old and the new by building on the vegetation heritage.

Giving residents such ample and generous spaces helps to recover the original intentions of social housing districts from the late 1960s: the quality of life that open space was trying to bring.

上图：南特，布雷尔马勒维勒街区
右图：勒蒙，格罗尼耶尔街区
右页上图：热纳维利埃，路德街区

Above: Nantes, Breil Malville neighbourhood.
Right: Le Mans, Glonnières neighbourhood
Opposite page, top: Gennevilliers, Luth neighbourhood

为长期做准备

植物栽种的计划策略需要把植物成长的速度列入考量。即使资源局促有限,还是可以为长期做准备。例如,密集地种植会引导树木成直线生长以便寻求阳光,因此只需要日后再降低林木密度即可,无需提前以人为的介入来修整这些会自理的"森林"。

Prepare for the Long Term

The rapid growth of plants must be taken into account in any vegetation strategy. You can prepare for the long-term, even with limited resources. Dense planting means trees will rise in a straight line to seek the light. Tree density can be reduced later, without having needed to intervene earlier in these "forests" which take care of themselves.

在一栋社会住宅楼的脚下,种植在石堆中的尤加利树以极少的费用创造了一种新的氛围,脱离了原来的贫穷形象(圣纳泽尔,圣马可街区)

At the foot of a social housing block, eucalyptus planted in riprap has inexpensively created an atmosphere dissociated from the original poverty (Saint-Marc, Saint-Nazaire)

圣纳泽尔，1992年及20年后的橡树林街区：实际存在的橡木森林重新赋予这地区名符其实的特色

In Saint-Nazaire, the Chesnaie (Oak Grove) neighbourhood in 1992 and twenty years later: a real oak forest reclaims its identity

green fingers 亲自栽种

法国 勒普莱西马塞 / 自1995年起

raising trees, with passion
以 热 情 栽 培 树 木

出自于个人的探险欲望，杰拉尔·佩诺（街巷工作室创始人）买下了这座美丽却荒芜已久的19世纪农牧园林。这片60公顷的林地、由山谷和池塘组成的历史遗产，就位于昂热城乡区域的边缘上。繁茂的树林、充满阳光的草地、专门用来培育"马奇耶树园"的一些地块，凡此种种都呈现出炼金术般的奇妙氛围。

让这个基地重新投入生产是对经济平衡的一种寻求，同时也是一种维持这块土地生存、积极抵抗城市蔓延扩张的方式。

一种创造性的经济

这个庄园荒芜已达25年有余，然而这个荒芜的状态却成为它更新的资源：
- 将自然苗种和野生灌木丛融入原始的空间布局内
- 在老化的地块上重新植林
- 为已达数百年树龄的树木事先进行更替的准备

水是关键性的流动元素：整个水周期的调节是依照"马奇耶树园"的需要而定的。

Through a desire for personal adventure, Gérard Pénot took over this beautiful agro-pastoral park dating from the 19th century, which had fallen into disuse. It comprised 60 hectares of woodland, a historical heritage of valleys and ponds, at the edge of the city of Angers. There is an alchemy expressed here, between wooded areas, meadows flooded with light, and plots reserved for growing Marcillé Trees.

Atelier Ruelle set up premises here and the area became a source of inspiration – through the richness of forms and colours, and by comparing how things were made with the requirements of forest management without public money.

To put the site back into use requires finding an economic equilibrium, which is also a way to keep the area intact, to actively resist urban sprawl.

A creative economy

The area was abandoned for 25 years and this neglect has become a resource for renewal:

- Natural seedlings and spontaneous thickets to be integrated into the initial composition.
- Ageing plots requiring reforesting.
- Anticipating the replacement of multi-century subjects.

Water is a critical carrier: its entire cycle is regulated to meet the needs of the Marcillé Trees.

培育这些树木加强了街巷工作室致力于土地整治规划的信念。动植物群落的现状对此地树木的分类管理产生影响，而公园则成为一种多元化经济的载体。面对时下对于乡村与都市、农业与郊区发展之间的关系所提出的质疑，此多元化经济的形态就是一个典范。

- 将草地保留作为牛群养殖的用途
- 就地割草为饲料
- 有限度地开发燃木和屋架木材（橡树、桦木、栗树、刺槐、甜樱桃树）
- 有实用性的公园能够创造就业机会和住宅

这个公园代表一种边界经济，它充满创造力，能够将农业活动以及与都市相关的活动（城市规划、建筑、景观）交织在一起。

Cultivating these trees deepened Atelier Ruelle's engagement in planning. Flora and fauna influence the management of tree collections and the park supports a diversified economy. This is an example of the sorts of questions being posed on the relationship between town and country, and between agriculture and suburban development.

- Meadows reserved for cattle.
- Cheap fodder
- The restrained exploitation of firewood and timber (oak, hornbeam, chestnut, locust tree, cherry tree).
- Jobs and housing in an inhabited park.

The park represents a border economy, creative in its ability to blend agricultural activities and activities related to cities – urban planning, architecture and landscape.

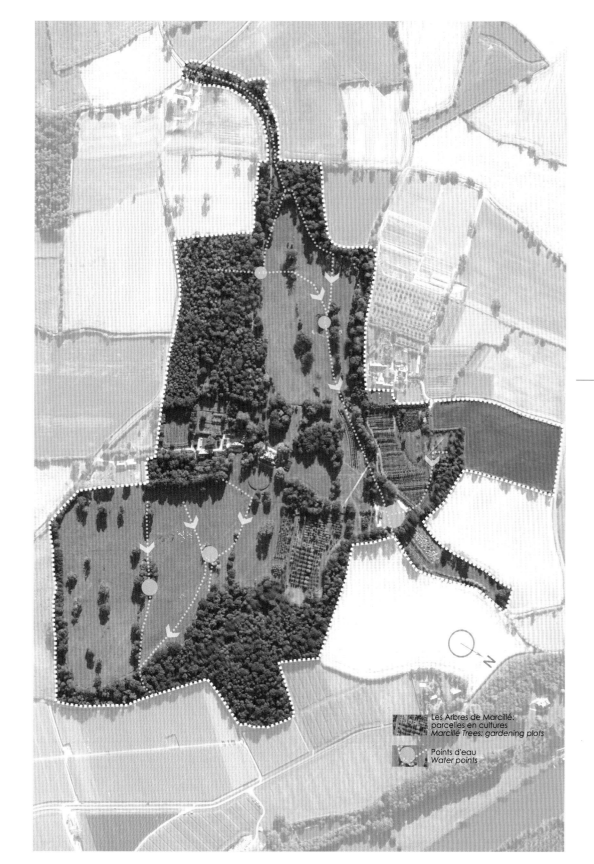

Les Arbres de Marcillé:
parcelles en cultures
Marcillé Trees: gardening plots

Points d'eau
Water points

观察是创造力的泉源

自由的形体、雕塑、建筑…… 对公园的观察激发人在对树木的规划选择上也展现相同的创造力，为树干和枝叶颜色进行搭配，并且塑造不同的体量和植物结构：直线排列或成丛栽种、林荫密集或林间空地……

Observation, the source of creativity

Free forms, sculpture, architecture... Observing the park encourages much creativity inspired by trees. To play with the colours of the trunks and foliage, building volumes and plant architecture, alignments or thickets, clearings or shade.

方 案 索 引
projects index

本书中介绍的项目
projects presented in this book

pp.12-25 Saint-Nazaire 圣纳泽尔
圣马克海滩、城市中心、火车站广场、布列特里街区、橡树林街区，1987-2011
城市-港口，2003-2012
BEACH OF SAINT-MARC, CITY CENTRE, STATION SQUARE,
BOULETTERIE NEIGHBOURHOOD, CHESNAIE NEIGHBOURHOOD, 1987-2011
CITY-PORT II, 2003-2012

合作者 / With
Technical Services of Saint-Nazaire City Council

业主 / For
Saint-Nazaire City Council, CARENE

pp.26-41 Nantes 南特
欧洲南特和马拉科夫车站 2001-2017
EURONANTES STATION AND MALAKOFF, 2001-2017

合作者 / With
Océanis BET VRD-Infrastructure

业主 / For
Nantes Métropole, Nantes Métropole Aménagement

pp.42-51 Rennes 雷恩
肯尼迪楼板平台步行区和维勒让街区1997-2007
KENNEDY DALLE AND VILLEJEAN NEIGHBOURHOOD, 1997-2007

合作者 / With
IOSIS BET VRD-infrastructure

业主 / For
Territoires Publics, Rennes City Council, Espacil Habitat

圣女贞德广场与乔治·贝尔纳诺斯广场，2005
JEANNE D'ARC SQUARE AND BERNANOS SQUARE, 2005

合作者 / With
IOSIS BET VRD-infrastructure

业主 / For
Rennes City Council

"蓝色计划"与公共空间规划纲要，1999-2001
BLUE PLAN AND PLANNING FRAMEWORK FOR PUBLIC SPACES, 1999-2001

业主 / For
Rennes City Council

圣马丁草原，设计竞赛2012
SAINT-MARTIN MEADOWS, COMPETITION 2012

合作者 / With
SAFEGE BET VRD-infrastructure-hydraulique, Ouest'AM BET environnement

业主 / For
Rennes City Council

pp.52-53 Saint-Dizier 圣迪吉耶
"绿树林"街区更新方案，2001-2013
VERT-BOIS (GREEN-WOOD) URBAN RENEWAL PROJECT, 2001-2013

合作者 / With
Technical Services of Saint-Dizier City Council

业主 / For
Saint-Dizier City Council

pp.54-55 Orly 奥利
大群体社会住宅区更新方案，2002-2014
GRAND-ENSEMBLE URBAN RENEWAL PROJECT, 2002-2014

合作者 / With
TETRA Programmation urbaine, Synerg CTS BET VRD-infrastructure,
Technical Services of Orly City Council

业主 / For
Valophis Expansiel, Orly City Council

pp.60-65 Veules-les-Roses 维勒雷洛斯
海岸区整治，1999-2013
WATERFRONT, 1999-2013

合作者 / With
Synerg CTS BET VRD-infrastructure

业主 / For
Veules-les-Roses City Council

pp.66-75 Lyon 里昂
多芬纳广场 2003-2005
DAUPHINÉ ESPLANADE, 2003-2005

合作者 / With
SEITT BET VRD-infrastructure

业主 / For
Urbane Community of Lyon

pp.76-83 Pontoise & Saint-Ouen l'Aumône 蓬图瓦兹&圣乌昂洛莫纳

阿兹河岸，2004-2011
QUAYS OF THE OISE, 2004-2011

合作者 / With
BATT BET VRD-infrastructure,
ASCODE E.Büschi Transport Engineering

业主 / For
Community of Agglomeration of Cergy-Pontoise

pp 84-89 Paris 巴黎

文森门大道，2005-2007
PORTE DE VINCENNES AVENUE, 2005-2007

合作者 / With
BATT BET VRD-infrastructure,
ASCODE E.Büschi Transport Engineering

业主 / For
Paris City Council

pp 90-93 Dijon 第戎

大学校园与伊拉斯谟广场，
第一阶段 2012年完成，第二阶段施工中
ÉRASME ESPLANADE, UNIVERSITY CAMPUS, 1ST PHASE 2012 AND
2ND PHASE IN PROGRESS

合作者 /With
Egis Aménagement BET VRD-Infrastructure (project representive),
Géodice E.Büschi Transport Engineering

业主 / For
Grand Dijon

pp 94-97 Saint-Étienne 圣艾蒂安

沙陀克高铁车站街区之公共空间，2008-2017
PUBLICS SPACES OF TGV CHÂTEAUCREUX STATION NEIGHBOURHOOD, 2008-2017

合作者 /With
SEITT BET VRD-infrastructure

业主 /For
EPA Saint-Étienne

pp 102-117

Angers 昂热

勾贝尔街区，2006
GAUBERT NEIGHBOURHOOD, 2006

合作者 / With
ACI BET VRD-infrastructures

业主 / For
Angers Habitat

玫瑰园街区，2004-2013
THE ROSERAIE (ROSE GARDEN) NEIGHBOURHOOD, 2004-2013

合作者 / With
SCE BET VRD-infrastructures

业主 / For
Society for the Development of the Region Angevine

Gennevilliers 热纳维利埃

路德街区 2002
LUTH NEIGHBOURHOOD, 2002

合作者 / With
BERIM BET VRD-infrastructures

业主 / For
Gennevilliers City Council

Le Mans 勒蒙

格罗尼耶尔街区，2012
GLONNIÈRES NEIGHBOURHOOD, 2012

合作者 / With
Cabinet Bourgois BET VRD-infrastructures

业主 / For
Le Mans Métropole

Nantes 南特

布雷尔马勒维勒街区，2006
BREIL MALVILLE NEIGHBOURHOOD, 2006

合作者 / With
SCE BET VRD-infrastructures

业主 / For
Nantes Métropole, Nantes Aménagement

马拉科夫街区更新，2001-2017
MALAKOFF NEIGHBOURHOOD URBAN RENEWAL PROJECT, 2001-2017

合作者 / With
Océanis BET VRD-infrastructures

业主 / For
Nantes Métropole

Saint-Denis 圣丹尼

儿童权利广场，2002
SQUARE OF THE RIGHTS OF THE CHILD, 2002

合作者 / With
Synerg BET VRD-infrastructures

业主 / For
Saint-Denis City Council, Plaine Commune Développement

Saint-Nazaire 圣纳泽尔

圣马克街区 与谢内（橡树林）街区，1996
SAINT-MARC NEIGHBOURHOOD AND THE CHESNAIE (OAK GROVE)
NEIGHBOURHOOD, 1996

合作者 / With
Technical Services of Saint-Nazaire Citi Council – BET VRD-infrastructures

业主 / For
Saint-Nazaire City Council

pp 120-129 Le Plessis-Macé 勒普莱西马塞

"马奇耶树园"，自 1995 年起
MARCILLÉ TREES, SINCE 1995

合作者 / With
Jean-Michel Guillier forest expert

业主 / For
SCI du Hêtre

其他项目 施工中或近期完成
other projects ongoing or recently completed

Garges-lès-Gonesse 嘎尔日-雷-戈内斯
北方白夫人街区城市设计
URBAN PROJECT FOR THE DAME BLANCHE NORD NEIGHBOURHOOD

合作者 / With
BERIM BET VRD, La Calade & AGI2D environment

业主 / For
Garges-lès-Gonesse City Council

Bourges 布尔吉
前波登军事医院基地改造计划与公共空间设计
PUBLIC SPACES AND URBAN PROJECT FOR THE SITE OF THE FORMER BAUDENS MILITARY HOSPITAL

业主 / For
SEM Territoria

Herbignac 埃尔比尼亚克
蓬巴与科尔杰斯汀基地公共空间与城市设计
PUBLIC SPACES AND URBAN PROJECT FOR THE POMPAS AND KERGESTIN SITES

合作者 / With
Egis BET VRD, Ouest'Am hydrology, BioTop environment

业主 / For
SELA

Champigny-sur-Marne 马恩河畔尚皮尼
莫尔达克街区城市设计
URBAN PROJECT FOR THE MORDACS NEIGHBOURHOOD

合作者 / With
GERAU Conseil urban planning

业主 / For
Champigny-sur-Marne City Council

La Courneuve 拉古尔纳夫
"北4000住宅"街区城市设计
URBAN PROJECT FOR THE "4000 NORD" NEIGHBOURHOOD

合作者 / With
BERIM BET VRD

业主 / For
Plaine Commune

Cholet 绍莱
让·莫内街区公共空间与住宅空间整治
PUBLIC AND RESIDENTIAL SPACES FOR THE JEAN MONNET NEIGHBOURHOOD

合作者 / With
IOSIS Grand Ouest BET VRD

业主 / For
Cholet City Council

Les Mureaux 雷米洛
"雷珀菲尔" 协议开发区城市设计与公共空间
URBAN PROJECT AND PUBLIC SPACES FOR "LES PROFILS" ZAC

合作者 / With
Franck Boutté Consultants environmental design and engineering, OTCI BET VRD-infrastructure

业主 / For
EPAMSA

Créteil 克雷泰伊
坡体佩-萨布里耶尔（小草原沙坑）街区公共空间与城市设计
PUBLIC SPACES AND URBAN PROJECT FOR THE PETIT PRÉ - SABLIÈRE NEIGHOURHOOD

合作者 / With
BERIM BET VRD, AGI2D environment

业主 / For
EXPANSIEL

Lille 里尔
宫科尔德-福布尔贝休恩街区城市设计
URBAN PROJECT FOR THE CONCORDE-FAUBOURG DE BÉTHUNE NEIGHBOURHOOD

合作者 / With
GERAU Conseil urban planning, Technicité BET VRD, Biotop environment

业主 / For
Lille City Council, LMCU, LMH

Lyon 里昂

汇流街区，佩拉什-圣布朗丹城市设计
URBAN PROJECT FOR PERRACHE-
SAINTE BLANDINE, LA CONFLUENCE

业主 / For
SPLA Lyon-Confluence

Reims 兰斯

红十字 街区，法国平原和艾森豪威尔区段
城市设计
URBAN PROJECT FOR THE CROIX
ROUGE NEIGHBOURHOOD – PAYS DE
FRANCE AND EISENHOWER SECTOR

合作者 / With
IOSIS Grand Est BETVRD

业主 / For
SA HLM Le Foyer Rémois, Reims Habitat

Lyon 里昂

汇流街区，第二协议开发区公共空间
PUBLIC SPACES FOR THE ZAC2, LA
CONFLUENCE

合作者 / With
Artélia BET VRD (project representive),
Agnès Deldon landscape designer,
L'Acte Lumière lighting designer, Acer
Campestre ecologist

业主 / For
SPLA Lyon-Confluence

Rennes 雷恩

莫勒帕-盖耶勒街区公共空间
PUBLIC SPACES FOR THE MAUREPAS-
GAYEULLE NEIGHBOURHOOD

合作者 / With
Arcadis BET VRD, Luminocité lighting
designer, Sonig energy

业主 / For
Territoires Publics, Rennes City Council

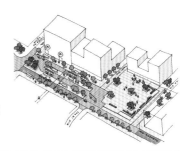

Massy 马西

马西-歌剧院街区，勃艮第-朗格多克协议开
发区公共空间与城市设计
PUBLIC SPACES AND URBAN PROJECT
FOR THE MASSY-OPÉRA NEIGHBOURHOOD
– BOURGOGNE-LANGUEDOC

合作者 / With
Technicité BET VRD

业主 / For
Massy City Council, SEM 92

Thionville 泰昂维

高处海岸街区城市设计
URBAN PROJECT FOR THE CÔTE DU
HAUT NEIGHBOURHOOD

业主 / For
Thionville City Council

Nanterre 南特尔

塞纳凯旋门街区，共和区段城市设计
URBAN PROJECT FOR THE SEINE ARCHE
NEIGHBOURHOOD – RÉPUBLIQUE SECTOR

合作者 / With
Arcadis BET VRD-acoustique-infrastructures

业主 / For
EPADESA

Treillières, Grandchamps, Nort-sur-Erdre
特雷利耶尔、格朗尚、埃尔德尔河畔诺尔
城市设计 PROJETS URBAINS
URBAN PROJECTS

合作者 / With
Egis BET VRD, Ouest'Am environment

业主 / For
Towns Community of Erdre and
Gevres, SAMOA (AMO)

Paris 巴黎

贝迪耶-布特胡协议开发区城市设计
URBAN PROJECT FOR THE BÉDIER-
BOUTROUX ZAC

业主 / For
SEMAPA

Villepinte 维勒班, 2011

前苗圃基地改造为生态街区之项目研究
PROJECT STUDY FOR AN ECO-NEIGHBOURHOOD
ON THE SITE OF LA PÉPINIÈRE

合作者 / With
OTCI BET VRD, Sétec Org urban
planning, AEU urban ecology, Franck
Boutté Consultants

业主 / For
Villepinte City Council

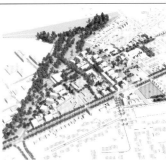

工作室简介

biography

杰拉尔·佩诺在完成城市规划方面的学业之后即成立了自己的事务所，并且同时修习建筑，也很快对公共空间产生强烈的热忱与关注。稍后，他与建筑师阿兰·傅尼耶在一个开放给多种领域学科的工作室合伙工作了几年。

1980年代，他为特拉普和圣纳泽尔两个当时情况艰难的城市进行城市设计，加强了他的信念：城市方案不仅是创作行为，也是对经济平衡的思考。因此，他以节约资源、扩大项目效益、关注日常生活空间的质量为重心，展开了与这两个城市的长期合作。

秉持着相同的精神，他带领着街巷工作室伴随圣迪吉耶、南特、昂热、奥利、勒蒙、雷恩等城市进行城市更新，也为巴黎、里昂、圣艾蒂安等城市构思方案。这个信念也延伸到"马奇耶树园"的经验上，将公园的诗意与栽植树木的热情结合，并且赋予这个位于昂热城乡区域边缘的场所一个崭新的经济模式。

After studying urban planning, Gérard Pénot set up his agency whilst he was still studying architecture, and soon found a passion for public space. He partnered with architect Alain Fournier for several years, in a workshop open to a variety of different professions, training courses and outlooks.

In the 1980s, he worked in Trappes and Saint-Nazaire, two cities in difficulty, where he forged the conviction that the project is as much about economics as it is an act of creation. By making the best use of resources, from the widest possible dissemination of the project to attention to the quality of everyday spaces, he has enjoyed many years of collaboration with these two cities.

It is always this approach that inspires the Atelier's work on the urban renewal of Saint-Dizier, Nantes, Angers, Orly, Le Mans and Rennes, as well as projects in Paris, Lyon and Saint-Étienne. It can also be found in the experience of the Marcillé Trees, which combines the poetry of the park and the passion of growing trees while rethinking the economy of a place at the border of the town of Angers.

credits 版权说明

撰文：弗蕾德里克·德·格拉维廉

照片与图片：街巷工作室

以下照片与图片除外：
Agence d'urbanisme de la région nazairienne – p.14, p.18下
Magdeleine Bonnamour – p.10, p.58, p.78上, pp.80-83, p.90, pp.92-93, pp.105-106, p.109, p.115下, p.122, p.124上, p.127, p.129
Chantier Graphique / Colas Vienne – p.51上, p.52, p.55上, p.64上 p.65上, p.69上, p.70, p.85上, p.96上
Xavier Depaule & Associés - p.85下, p.88下
Gérard Dufresne – p.4, pp.12-13, p.15, p.17, pp.19-25, p.27, p.31, p.33下, pp.38-39, p.40下, p.41下, p.42, p.48下, p.49, pp.60-63, pp.76-77, p.79, p.84, p.86 haut, p.87下, p.110下, p.111上, p.117上
Zoé Fontaine – p.51下
Valery Joncheray – p.26
Philippe Schuller – p.95下
Service de communication Ville de Saint-Dizier / Erick Colin – p.53
Territoires – p.43, p.44下
Valophis Habitat – p.54, p.55下

Texts: Frédérique de Gravelaine

Illustrations and photographies: Atelier Ruelle

except the following photographies and illustrations:
Agence d'urbanisme de la région nazairienne – p.14, p.18 bottom
Magdeleine Bonnamour – p.10, p.58, p.78 top, pp.80-83, p.90, pp.92-93, pp.105-106, p.109, p.115 bottom, p.122, p.124 top, p.127, p.129
Chantier Graphique / Colas Vienne – p.51 top, p.52, p.55 top, p.64 top, p.65 top, p.69 top, p.70, p.85 top, p.96 top
Xavier Depaule & Associés - p.85 bottom, p.88 bottom
Gérard Dufresne – p.4, pp.12-13, p.15, p.17, pp.19-25, p.27, p.31, p.33 bottom, pp.38-39, p.40 bottom, p.41 bottom, p.42, p.48 bottom, p.49, pp.60-63, pp.76-77, p.79, p.84, p.86 top, p.87 bottom, p.110 bottom, p.111 top, p.117 top
Zoé Fontaine – p.51 bottom
Valery Joncheray – p.26
Philippe Schuller – p.95 bottom
Service de communication Ville de Saint-Dizier / Erick Colin – p.53
Territoires – p.43, p.44 bottom
Valophis Habitat – p.54, p.55 bottom

contributions 致谢

2013年工作团队：

Thomas Audouard, Olivier Delbano, Rémi Deverrewaere, Élisabeth Georges, Malika Goudard, Thibault Groussin, Gaëlle Guibert, Emmanuel Jaulin, Julia Kapp, Carine Lassalle, Philippe Lebrun, Anne Margairaz, Charlotte Moisand, Olivia Moyse, Véronique Navet, Gérard Pénot, Philippe Pinson, Alexandra Poncet, Oualid Satouri, Camille Taque, Colas Vienne

以往职员：

Yohann Chansellé, Éric Milet, Isabelle Moulin, Jean-François Poisson, Yann Renoul, Guillaume Sevin, Clarel Zéphir

感谢Dominique Doan和Luce Pénot对本书编辑的协助。

Team in 2013 :

Thomas Audouard, Olivier Delbano, Rémi Deverrewaere, Élisabeth Georges, Malika Goudard, Thibault Groussin, Gaëlle Guibert, Emmanuel Jaulin, Julia Kapp, Carine Lassalle, Philippe Lebrun, Anne Margairaz, Charlotte Moisand, Olivia Moyse, Véronique Navet, Gérard Pénot, Philippe Pinson, Alexandra Poncet, Oualid Satouri, Camille Taque, Colas Vienne

Former employees :

Yohann Chansellé, Éric Milet, Isabelle Moulin, Jean-François Poisson, Yann Renoul, Guillaume Sevin, Clarel Zéphir

Thanks to Dominique Doan and Luce Pénot for their assistance in the preparation of this book.

图书在版编目（CIP）数据

对谈：街巷工作室设计作品专辑 /（法）弗蕾德里克·德·格拉维廉撰文. -- 沈阳：辽宁科学技术出版社，2013.8
ISBN 978-7-5381-8152-4

Ⅰ. ①对… Ⅱ. ①弗… Ⅲ. ①城市规划－建筑设计－作品集－法国－现代 Ⅳ. ①TU984.565

中国版本图书馆 CIP 数据核字(2013)第 154911 号

出版发行：辽宁科学技术出版社
　　　　　（地址：沈阳市和平区十一纬路29号 邮编：110003）
印 刷 者：利丰雅高印刷（深圳）有限公司
经 销 者：各地新华书店
幅面尺寸：210mm×230mm
印　　张：8.5
插　　页：4
字　　数：50千字
印　　数：1～1500
出版时间：2013年 8 月第 1 版
印刷时间：2013年 8 月第 1 次印刷
责任编辑：陈慈良　隋　敏
封面设计：亦西文化
版式设计：亦西文化
责任校对：周　文
书　　号：ISBN 978-7-5381-8152-4
定　　价：88.00元

联系电话：024-23284360
邮购热线：024-23284502
E-mail: lnkjc@126.com
http://www.lnkj.com.cn
本书网址：www.lnkj.cn/uri.sh/8152